Corporate Life in the Digital Music Industry

Alternate Takes: Critical Responses to Popular Music is a series that aims to examine popular music from critical perspectives that challenge the accepted ways of thinking about popular music in areas such as popular music history, popular music analysis, the music industry, and the popular music canon. The series ultimately aims to have readers listen to – and think about – popular music in new ways.

Series Editors: Matt Brennan and Simon Frith

Editorial Board: Daphne Brooks, Oliver Wang, Susan Fast, Ann Powers, Tracey Thorn, Eric Weisbard, Sarah Hill, Marcus O'Dair

Other Volumes in the Series:

When Genres Collide by Matt Brennan

Nothing Has Been Done Before: Seeking the New in 21st-Century American Popular Music by Robert Loss

Annoying Music in Everyday Life by Felipe Trotta

DIY Music and the Politics of Social Media by Ellis Jones

Owning the Masters: A History of Sound Recording Copyright by Richard Osborne

Dance Music: A Feminist Account of Ordinary Culture by Tami Gadir

Corporate Life in the Digital Music Industry: Remaking the Major Record Label from the Inside Out by Toby Bennett

A Musical History of Digital Startup Culture by Cherie Hu (forthcoming)

Live from the Other Side of Nowhere: Contemplating Musical Performance in an Age of Virtual Reality by Sam Cleeve (forthcoming)

National Phonography: Field Recording, Sound Archiving, and Producing the Nation in Music by Tom Western (forthcoming)

Corporate Life in the Digital Music Industry

Remaking the Major Record Label from the Inside Out

Toby Bennett

BLOOMSBURY ACADEMIC
NEW YORK • LONDON • OXFORD • NEW DELHI • SYDNEY

BLOOMSBURY ACADEMIC
Bloomsbury Publishing Inc
1385 Broadway, New York, NY 10018, USA
50 Bedford Square, London, WC1B 3DP, UK
29 Earlsfort Terrace, Dublin 2, Ireland

BLOOMSBURY, BLOOMSBURY ACADEMIC and the Diana logo are trademarks of
Bloomsbury Publishing Plc

First published in the United States of America 2024

Copyright © Toby Bennett, 2024

For legal purposes the Acknowledgements on p. viii constitute an extension of this
copyright page.

Cover design by Louise Dugdale
Cover images: Top, zamrznutitonovi/Getty Images, Bottom, mnirat/Adobe Stock

A catalog record for this book is available from the Library of Congress.

ISBN: HB: 978-1-5013-8723-4
PB: 978-1-5013-8722-7
ePDF: 978-1-5013-8725-8
eBook: 978-1-5013-8724-1

Typeset by Deanta Global Publishing Services, Chennai, India
Printed and bound in Great Britain

Series: Alternate Takes

To find out more about our authors and books visit www.bloomsbury.com and
sign up for our newsletters.

Contents

List of Acronyms   vi
Acknowledgements   viii

Introduction   1

Part I Outside in

1   Working for the man?   9
2   Access all areas   30
3   Big Music   53

Part II Inside out

4   Re-evaluation   79
5   Passion work   95
6   Standardization   119
7   Systems work   130
8   Professionalization   150
9   Knowledge work   168

Conclusion: Everyone's a critic   191

Appendices   205
Notes   211
Bibliography   215
Index   242

Acronyms

A&R: Artists and Repertoire

AIM: Association of Independent Music

BECTU: Broadcasting, Entertainment, Cinematograph and Theatre Union

BMG: Bertelsmann Music Group

BPI: British Phonographic Industry

CBI: Confederation of British Industry

CSR: Corporate Social Responsibility

DCMS: Department for Culture, Media and Sport

EMI: Electric and Musical Industries

FAC: Featured Artists Coalition

FE: Further Education

HE: Higher Education

IFPI: International Federation of Phonographic Industries

ISRC: International Standard Recording Code

ISWC: International Standard Work Code

MAP: Music Academic Partnership (UK Music)

MCPS: Mechanical-Copyright Protection Society

MMF: Music Managers Forum

MPA: Music Publishers Association

MPG: Music Producers' Guild

MU: Musicians' Union

PMS: Popular Music Studies

PPD: Published Price to Dealer

PPL: Phonographic Performance Ltd

PRS: Performing Rights Society

SME: Sony Music Entertainment

UMG: Universal Music Group

UPC: Universal Product Code

WMG: Warner Music Group

Acknowledgements

This book has been on a long journey. Enormous thanks are due to all those who donated their time, explained their justifications and offloaded their apprehensions to me in interviews and other conversations – but who must remain anonymous. I can, however, mention the generous assistance in arranging contacts from Ben Duke (Sony), Ray Pope (Warners) and Sarah Palmer (Universal) in the UK, as well as Amanda Pede in Santa Monica. At UK Music, Cathy Koester, Jonathan Todd, Lorna Finlayson, Oli Morris and Natalie Williams were hugely supportive, while Alison Wenham, Andy Edwards and Robert Ashcroft gave valuable feedback on the report that fed into Chapter 8. Julia Payne of The Hub was a source of further access and later invaluable mentorship. Arguably this all began when I arrived at High Street Ken tube for my first proper London job back in 2007 – I thank Sheenagh James and Kevin Phelan for agreeing that I might be a useful addition to the team. Big shout out to the original licensing crew, especially Pete Thompson, Emine Rifat, Lee Puddefoot, Anna Dueweke, Kate Bestic . . . and everyone else!

I am grateful to Andy Pratt for his early support and guidance in the original research and his critical mentorship ever since. He, Sean Nixon, and Chris Bilton encouraged me to think there might be a book there – with only a few tweaks . . . At KCL, I acknowledge the generous funding support from the Graduate School (without which nothing); as well as what was then a small but close-knit academic community in CMCI – in particular, Nick Wilson's intellectual and administrative guidance remains with me, as well as conversations with Ben Rampton, Bridget Conor, Giuseppe Zevolli, Laura Speers, Mark Dawson, Natalie Wreyford, Paula Serafini, Rachel O'Neill, Sara de Benedictis and Simon Hewitt. The Cultural Institute (as it then was) supported the work with UK Music and The Hub; and the counselling service supported me as a human when I didn't know how much I needed it.

Since finishing the initial research, the manuscript has gone through several development phases. Thanks to the gift of a series of fixed-term contracts, my thinking has benefited from the ideas and academic companionship at various stages of Alex Reynolds, Amin Samman, Andre Spicer, Anubha Sarkar, Bailey Adie, Bill Acharjee, Chris Anderton, Dan Ashton, Diana Yeh, George Musgrave,

Hannah Curran-Troop, James Hannam, Jamie Reddington, Jane Parry, Jenny Mbaye, Jo Littler, Julia Toppin, Kim-Marie Spence, Martin James, Nick Seaver, Paromita Saha, Paul Rutter, Ros Gill, Sareata Ginda, Si Long Chan, Tarek Virani and Xanthia Mavraki, among many others. I was lucky to spend several years working on two European projects: the CICERONE team forced me to think more carefully about space and value chains (see Chapter 3), and the CIRCE team made me do the same for culture, crisis and 'diversity' (see the Conclusion). My good friend Jonny Stafford was a persistent interlocutor who made me think about Marx and pirates (the seafaring kind). Liz McFall, Darren Umney and Philip Roscoe at JCE made me want to do something 'useful' as well as 'critical'. I have been incredibly fortunate to work with a brilliant teaching team at the University of Westminster – in particular, I need to thank the inimitable and irrepressible Sally Anne Gross, who pushed me to make ideas matter to people. Sally and Paul Rekret read drafts of the final manuscript and prodded me in the right direction where needed.

I am indebted to the generosity of Simon Frith and Matt Brennan, who agreed that a book was in there somewhere, and several anonymous reviewers whose close attention helped me to excavate it. Leah Babb-Rosenfeld and Rachel Moore at Bloomsbury have stewarded it through shifting deadlines to completion and translated my hunch that record boxes and filing cabinets might be a fun visual pun into something that somehow works as a front cover.

So – a long journey. Along the way, I lost my mum, my dad and my big brother. They all left their mark here too: for better or worse, they supported my own 'passion for music', and basically every decision I made as a result, unconditionally. So this book is dedicated to them, as well as to my old friend Robb Skipper, who was far too young. He 'generated the music that makes you feel better'.

Anna's been there all the way from the beginning. Thank you for all your patience and support. Erik turned up halfway through, and Maja a few years later. Please don't either of you pursue a career in music – but if you must, read this first.

Parts of the book have originally been published in a different form in other contexts:

Short sections of Chapter 5 appeared in: Bennett, Toby (2018a), '"Essential – Passion for Music": Affirming, Critiquing and Practising Passionate Work in Creative Industries', in Lee Martin and Nick Wilson (eds.), *The Palgrave Handbook of Creativity at Work* (London: Palgrave), 431–459. Reproduced with permission of Palgrave Macmillan. DOI: 10.1007/978-3-319-77350-6_21

Sections of Chapters 6 and 7 are derived in part from: Bennett, Toby (2020a), 'Towards "Embedded non-creative Work": Administration, Digitisation and the Recorded Music Industry', *International Journal of Cultural Policy* 26:2, 223-238. Copyright Taylor & Francis. DOI: 10.1080/10286632.2018.1479399

Chapter 8 is based on: Bennett, Toby (2015), *Learning the Music Business: Evaluating the "Vocational" Turn in Music Industry Education* (London: UK Music)

A section of the concluding chapter is based on: Bennett, Toby (2018b), '"The Whole Feminist-Taking-Your-Clothes-Off Thing": Negotiating the Critique of Gender Inequality in UK Music Industries', *IASPM@Journal* 8:1, 24-41. Reproduced under a Creative Commons licence with permission of IASPM. DOI: 10.5429/2079-3871(2018)v8i1.4en

Introduction

How is a media industry remade when faced by the rapid take-up of a technology it does not control? This book follows what happened inside major record companies – dominant firms in the global production, circulation, consumption and regulation of commercial popular music – after they experienced such a loss of control and asks how they came to regain it. There already exist many answers to this question. Those who lead these 'major labels' inevitably offer a personalized account, highlighting their artistic and business expertise, their empathic relationships with musicians, their inherent tenacity and capacity to innovate. More critical explanations typically centre on financial and political power: their ability to weather the storm, to buy up or sabotage competitors, invest in alternatives and to influence governments, markets and distribution channels around the world. Either way, music corporations remain attractive routes to market for many artists and creatives, even if, as many continue to believe, this involves making something like a pact with the devil. All of this is true enough, as far as it goes. But I will be adding another explanation, one which I think helps us grasp the multidimensional nature of this process. What I want to emphasize is that these organizations are made up of people; moreover, they are made up of people *at work*. This is common-sensical enough but the implications are not often enough taken on board by scholars, critics, creators or consumers of popular music. The evolution of a firm is not natural or inexorable. There is no change, and certainly no recovery, without the *work* that makes it happen. They can, perhaps, be remade differently again.

At the heart of my argument is the matter of how industrial transformation comes to be imbued with moral force. In particular, my focus is on the fate of the largest corporations at the centre of the British recorded music industry, over the period of 1999–2015. These dates are not arbitrary. The first marks the point at which revenues from sales of recorded music began to plummet, marking a 'significant break' with the previous 'golden era', in economic geographer Andrew Leyshon's (2014: 80–8) words, in which the record industry 'enjoyed about fifteen years of steady growth'. The second date is the moment when income from downloading and streaming outstripped that from physical sales, leading to steady growth once again (IFPI 2016) and a significant increase in the global workforce after years of staff cuts and stagnation (Ingham 2023a). Despite the pronouncements of 'disintermediation' that characterized the early 2000s (e.g. Jones 2002), the majors managed to maintain, even consolidate, their powerful grip on systems of music production. Quite reasonably, then, we might ask: what 'transformation'? Is this not just a return to business as usual? Perhaps. But my contention, outlined across these pages, is that the period was significant insofar as its outcome was not inevitable. Preserving existing power relations required new structures to be put in place, at the level of individual firms; it also produced new avenues of potential future change in these firms. But no transformation is possible, past, present or future, without the active consent, tacit or explicit, of those who worked in them.

It is easy to forget quite how much criticism was launched in those early years, from both inside and outside the industry – and how triumphant it was. Commentators told the story of disruptive disintermediation with considerable glee. 'Most of what's been written over the past ten years', wrote Fred Goodman (2010: 275) of the 2000s, makes it 'hard to imagine the major labels were anything more than criminal enterprises'. The boom period of the 1980s and 1990s had given them free rein to behave 'negligently and in ways that have served to emphasize their abuses of market power' (Burkart 2010: 4). For artist manager Simon Napier-Bell (2008), these companies thereby 'brought it on themselves', and their 'suffering' is simply the price they pay for past crimes. 'Informed consumers', on the other hand, have been shown to constitute a 'dangerous variable to an industry largely based on bullshit' (Barfe 2006: n.p.). Less Machiavellian accounts simply emphasized the 'complacency' that had set in, chastising majors for 'failing to recognize or grasp the opportunities' of digitization after having been 'careless and wasteful in their investment' (Rogers 2013: 55) – a diagnosis echoed by actors within the organizations themselves (Forde 2019; Wheeldon 2014). A strategy of taking those informed consumers to court, and in some cases to prison, hardly helped the publicity drive. From the late 2000s on, a new approach was found, one that focused on offering the public more convenient alternatives for accessing digital music, rather than fining them for doing so, while shifting regulatory pressure onto internet service providers and other network intermediaries. Thus, the anti-piracy refrains became less visibly associated with tumbling profits and punitive measures, instead taking a more instructional tone. Trade bodies emphasized that sustainable careers are built on authentic connections and just rewards, 'educating' fans that piracy undercuts their ability to forge meaningful relationships with the artists they love (Edwards et al. 2015).

In such ways, music's *crisis of reproduction*, as Leyshon (2014: 88) put it, plunged major labels into a parallel *crisis of legitimation*, to use the words of the critical theorist Jürgen Habermas (1975), requiring communicative action. This was not just a technical issue, it was not about manufacturing capacity or pricing, nor could it be solved by the top-down reassertion of oligopoly power alone. Given that the possibility of accessing music without payment had never before been a viable option, at least at scale, the need for legitimacy was quite new. A lack of legitimacy produces a crisis of leadership. Major labels increasingly required public assent to lead what they and others saw as necessary changes in the digital cultural economy: listeners needed to be convinced to part with their cash, while governments, as elected representatives of a national citizenship, needed to be convinced to renovate and reinforce a more comprehensive copyright regime. But behind the scenes, crucial organizational changes needed to be implemented, and, moreover, people were needed to enact them. This work would not always be glamorous or fun (though it could be, at times). Despite popular imagery, on a day-to-day level, it was commonly mundane, technical, routine, managerial and administrative work, involving glitchy software, spreadsheets, cabinets of contracts and weekly update meetings. This is not the kind of activity that fills the pages of magazine gossip columns or bestselling autobiographies – or, indeed, much in the way of academic scrutiny. Indeed, it is a common source of frustration for workers themselves. Yet it was crucial simply to keep companies running throughout this dual crisis. Employees' participation did not take place under conditions of extreme duress – but neither was it unconditional nor always enthusiastic. This book is about them. Giving life to corporate work, it asks: who participated in this transformation, what work did they do, under what conditions and contradictions – and how did they make sense of it all?

While drawing on the experiences of a range of actors, my narrative therefore centres the backstage world of non-creative mainstream music work, beyond the most visible artists, executives and entrepreneurs. It is based on an in-depth qualitative study (2012–16) of London's three major record label offices and the broader social world in which their employees were embedded – including continuous informal interaction outside the workplace, attendance at industry events, conferences and festivals, and immersion in trade journalism and business commentary. In addition to the many, many conversations I had with employees, I conducted twenty-three on-the-record but anonymous interviews with individuals connected in some way to at least one major label, typically several, with most having a varied 'portfolio' career.[1] They ranged in seniority from entry-level to C-Suite executive, via middle management, and included artist managers, analysts, educators and commentators outside the majors. This study built on my own previous career within a major label catalogue department, departing as a licencing manager (2007–12), at the point that this particular function was moving, in line with a digitising business model, from the organizational periphery to its core. This I augmented by another, more strictly ethnographic, six-month placement within a major (2013), recording observations while completing the tasks I had been set. Part of the research, then, involved constructing a reflective critical memoir on each period, elements of which inform and periodically intrude into the book's narrative. I was further enabled by two smaller projects (2014–15): conducting a report on music business and higher education for the national trade body UK Music; and advising a funding and support programme for digital music entrepreneurs, run by The Hub, a cultural 'think and do tank'. I am indebted to key individuals in these organizations for inviting me in, putting me in contact with helpful informants, and helping me to remain engaged with wider industry debates outside the core focus of the research. The book is based, then, on a multi-sited (auto-)ethnography, covering almost a decade of immersion in total.

The first part of the book consists of three chapters, moving 'outside in' to locate the argument within a particular time and space. Chapter 1, 'Working for the Man', sets the book's argument within a longer historical trajectory, placing the question of legitimacy within the contemporary corporate music workplace in the long trajectory of twentieth-century commercial popular music, with its anti-corporate, transgressive outsider sentiments. It does so in conversation with the sociologists Luc Boltanski and Ève Chiapello, who famously argue that the ongoing critique of capitalism works as a motor for its spiritual renewal. Chapter Two, 'Access All Areas', zooms into the core period under study, exploring the apparent epochal shift into an 'age of access', beginning in 1999. This is a moment, and a concept, that I argue demands interrogation: not only in terms of changing access to music consumption but equally in terms of labour markets (access to work) and control over information and expertise (access to knowledge). These three lenses – music consumption, work structures, knowledge production – establish the analytic terrain that structures the second half of the book. Finally,

Chapter 3 places that analysis in its specific empirical context. It sets out the scope of the major label as an employer, in terms of three companies occupying a particular set of buildings in a small corner of London, with a shared mode of production and distinctive professional culture. I introduce the fictional 'Sonaverse Records' – a hybrid version of the real majors – and consider the implications of the 'Generation X' executives moving into key decision-making positions during this time – and bringing a particular set of cultural values with them.

The subsequent six chapters comprise the second part of the book – this time working 'inside out', in the form of three sets of paired arguments: each pair first traces the organizational implications of a broad public or industry-wide shift; followed by a more detailed exploration of how this shift plays out from the perspective of working life. They move through the themes set out in Chapter 2. First, a *re-evaluation* of past sins took place, through media representations of music industry practice (in books like John Niven's *Kill Your Friends* and TV shows like *X Factor*) putting the performance of critical judgments on public display, coupled to internal developments in consumer 'insights' and the management of fan-like *passion* at work (Chapters 4 and 5). Second, the *standardization* of a digital infrastructure enabled music to be 'unbundled' from the sale of physical goods, enabling the capture of data and ongoing rent-like micro-payments, but requiring an expansion of routine *systems* work (Chapters 6 and 7). Third, a contested *professionalization* of music careers took shape through the expansion of higher education, alongside a growth in expert *knowledge* work, with the ambition to curate ideas and train a future workforce in a format suited to industry's needs (Chapters 8 and 9). Each of these paired chapters tracks twin processes of industrial rationalization and the continuous reiteration of the 'passion for music' (Bennett 2018a), together documenting what I refer to in the final chapter as *passionalization*. I conclude by considering how internal and external critiques have been launched to address the most recent iterations of the digital music industry, since the core era of the study. Within and against these processes, I want to surface what appears to be a persistent ethic of cultural office work, which I refer to as *meaningful maintenance*. Rethinking the major label in this way, from 'inside out', I argue that it is worth paying close attention to such work for its capacity to shed light on how cultural-industrial change does or does not take place – and, moreover, how it might take place otherwise.

Part I

Outside in

Working for the man?

Buskers, Beatles and record men

It's August 2009. Writing his regular column for *The Guardian* newspaper, John Harris is considering a government proposal, supported by the record company trade association British Phonographic Industry (BPI), to disconnect broadband connections for illegal file-sharers.

> *Given the popularity of illicitly distributing music, films, games and software, are internet service providers really going to monitor and then shop so many of their own customers? Such are the drawbacks of an idea that [. . .] looks likely to have been firmly put on the agenda by Lucian Grainge, the chairman of the behemothic Universal Music, and an analogue man if ever there was one. (Harris 2009)*

After a series of high-profile court cases demonizing persistent file-sharers, this was just the latest in a suite of major label-driven draconian punishments and, Harris argued, soon to be outmoded by technological change. In the course of outlining the trials and tribulations of the new digital music landscape then taking shape, Harris took care to distance his scepticism towards top-down industrial change management from what he termed 'the great chaotic utopia envisaged by some online evangelists', which could only produce a 'culturally impoverished' future with 'millions of buskers, but no Beatles' (Harris 2009). The evocative dualism pitched a vision of the coming music industry in terms of two extremes. Do you really want, asks Harris, a world populated with amateur musicians scrabbling for pennies and the attention of a meagre few? Or do you want an iconic, epoch-defining band that transformed pop-rock from mere entertainment into a legitimate art form, routinely recognized as the pinnacle of the West's twentieth-century popular culture? This comparison presages much of the subsequent critical discourse on music streaming (cf. Hesmondhalgh 2022; Marshall 2015). As a former music critic, editor of *Select* magazine and documentarist of the Britpop movement that revived a public interest in home-grown guitar bands in the 1990s, we might be unsurprised to find out where Harris' own allegiances lie.

Despite mention of the 'behemothic' music corporation, the sharpest end of Harris' critique was reserved for Silicon Valley techno-utopians, the Swedish libertarian Pirate Party, and what he termed 'liberal-left' proponents of 'participatory production'. These had imbibed the 'Californian Ideology' (Barbrook and Cameron 1996) of the 1960s counterculture as it flowed into 1980s cyberculture, via the likes of Stewart Brand and The Grateful Dead's John Perry Barlow, with its communal ideals dialled down and its libertarian aspirations dialled up (Turner 2006). They had advocated for an electronic architecture that assumed music, like information, 'wanted to be free'. Such ideas flowed through MIT Media Lab's director Nicholas Negroponte (1996), who had proclaimed that digitization would lead to the 'disintegration' of copyright; *Wired* magazine editor Kevin Kelly (2002), who predicted that 'the recording industry as we know it is history' (cf. Rogers 2013: 8–9); and the influential lawyer, 'copyleft' reformer and 'free culture' advocate Lawrence Lessig. As well as e-commerce ideologues like *Wired*'s Chris Anderson, author of *Free! The Future of a Radical Price*, Harris singled out Swedish filesharing services The Pirate Bay, whose founders had been imprisoned earlier that year in a highly-publicized case that had led to a growth in membership of the Pirate Party and election to the European parliament. In passing, he mentioned the up-start Spotify. Recently launched in Europe with the promise of a solution to piracy – shifting from a closed, unlicensed peer-to-peer filesharing protocol into a licenced platform, eligible to stream major label-owned content, with an iPhone app in the works – Spotify's business model and profit-making potential, in the wake of a global financial crisis, remained vague at best (cf. Eriksson et al. 2019). Cautiously embracing efforts to reform the terms and duration of copyright law, Harris ultimately aligned himself with the digital critics: the likes of Andrew Keen (2007: 7; 108), for instance, who had denounced a broader 'cult of the amateur', set to undermine established 'cultural standards' and 'moral values', and lamented the dismantling of expertise embodied in long-standing institutions – such as 'the increasingly desperate record industry'. 'There's also a case to be made for societal norms, and the responsibility of all of us to observe them', Harris opined, whether through supporting the artists you love with hard cash or teaching your kids that creativity deserves a reward: 'at the core of any sustainable culture, there's a crucial knot of ethics and pragmatism that the year-zero libertarians are set on undoing' (Harris 2009).

The comment on major label head Lucian Grainge – a mere 'analogue man' in a digital age – was a much briefer aside. Never one to let the record go uncorrected, however, a concise, carefully worded but forceful riposte from the CEO appeared in *The Guardian*'s letters page three days later (Grainge 2009). Taking issue with the epitaph, he pointed to the company's investments in digital platforms like iTunes and Spotify, and defended threats of internet suspension as 'tough measures' to 'create a level playing field' for 'content owners' in the name of 'British creativity'. In light of the latter, he endorsed Harris' 'artful' metaphor of many buskers and no Beatles, but, by contrast, it was Grainge himself who, by carefully stewarding capital in the right direction and standing up for artists when engaging with powerful players, would be the true saviour of 'the digital generation of talent'. Rather than an analogue man, Grainge implied he was simply a music man, or, more appropriately, a *record man*.

Typically used as a mark of respect – 'he was a *real record man*' – this is 'an old term, somewhat ambiguous but nonetheless an honor reserved for a certain caliber of entrepreneur who *knows* his music' (Murphy 2014: ix, emphasis in original). Originating in post-war New York and London, record men were fast-talking, countercultural non-conformists, typically Jewish emigrés (Karp 2015): experts in the art of taking the soul to market, schooled in small-scale enterprise, they were open to not a little glamour and debauchery. In the United States, in the journalist and author's Rich Cohen's words, the likes of Phil and Leonard Chess exemplified a group of record men in the 1940s and 1950s:

> *geared to the low rumble of the Negro street: because they believed there was money in it, because they liked the music, because it was fun, because no one could shut them out, because chicks, chicks, chicks, because make your own hours, because it's better than selling schmatas, because if you want something build it, and also because, as immigrants with a personal knowledge of persecution, they were maybe the only Americans willing to form a partnership with poor blacks.* (Cohen 2004: 56)

In the UK, the term was less widely used, but nonetheless it captures a similar origin story, centred on a coalition of hucksters and agents clustering around the theatre halls, sheet music publishers and espresso bars of late-1950s London. Larry Parnes is the exemplar here: 'the original model of the pop Svengali', for the historian Jonathan Karp (2015: 143), 'the impresario as creative artist whose imagination and gift for presentation constitutes the truly artistic component in the otherwise ephemeral and disposable genre of pop music fashion' – later emulated by the Beatles' Brian Epstein and the Rolling Stones' Andrew Loog Oldham, among others. Such figures were crucial in brokering what Karp refers to as a 'rock 'n' roll international' between the United States and UK scenes and markets, challenging the dominance of the major recording companies of the time. As rock 'n' roll turned into big business, later iterations of such figures would be responsible for keeping the ethic of independence alive. The journalist Gareth Murphy's history of indie music is told through the stories of such men, for instance. While attuned to the volatile passions of the crowd and the money-making potential that lay behind it, he argues that these enterprising outsiders were always 'motivated by something higher':

> *Behind every emblematic record label lies a hidden story – usually deep within the boy. These patrons of contemporary music are the A-students of the music business, who view their role as judge and guardian. (Murphy 2014: xii)*

Behind the scenes, in other words, the figure of the record man came to embody a deep-rooted sense of responsibility and cultural custodianship.

This chapter traces the story of how this professional 'ethic', understood in Max Weber's classical sense of a vocation or 'calling' (Weber 2001), came into being in the context of commercial music production – and how it came to be in need of renewal – by articulating a critique of work that lined up with a broader 'spirit of capitalism' (Boltanski and Chiapello 2005). In dialogue with this, I suggest here that the critical academic field of popular music studies (PMS) has often taken on and given expression to elements of this spirit, uniting both in a shared 'moral economy'. It is within the record men, according to Murphy and others like him, that the 'truth' of the music industry might be found. It is a claim that also characterizes much critical scholarship. For instance, Andrew Mall (2018: 445) forcefully argues that the academic field of popular music studies (PMS) has historically exhibited an 'indie prejudice', as he calls it, 'a dismissive approach to the study of major labels and musical mainstreams that impacts our ability, as a scholarly field, to speak with authority about the largest segments of the commercial record industries and thus the most popular of popular musics'. Whether we fully accept this provocation depends on the contested definition of 'most popular'. But if such a prejudice exists, it is apiece with a more widespread belief in the centrality of indie record men to why the industry might matter at all. It is such a belief that I want to contest here.

I first elaborate on what is meant by the entwinement of *moral values* and *economic value*, before turning to the confluence of post-war rational industrial production with the 1960s counterculture that finds one expression in Roy Orbison's 'Working for the Man'. With the critique of the factory, and later waged employment per se, comes the embrace of alternative, independent and creative endeavours that characterize subsequent transformations in the 'spirit of capitalism' – albeit with more ambivalence than is often supposed, especially with respect to gender. But if critical knowledge production has closely tracked the developing spirit of musical capitalism, it has led to a disavowal of the kind of work that the corporate Man still requires. By the end of the 2000s, many critics argued, this spirit was in danger of being lost entirely. As both Harris and Grainge suggest, the key site of struggle by this point was no longer viewed as that between 'indies' and 'majors' but the global battle to protect a record business caught up in an unfurling information-driven hyper-capitalism, dominated by a small handful of large extractive tech firms. The 'something higher' motivating these men to stake out respective positions as (in Murphy's words) 'judge and guardian' was the very idea of a music industry at all. I argue that one way to productively move beyond the impasse is to pay close attention to the mundane 'maintenance' work that sustains mainstream industry structures.

Value and values

We are already deep in perilously myth-laden terrain. Nonetheless, the question of how a socialist music critic comes to converge with the CEO of a corporate multinational on a common vision of the cultural future, centred on icons of white male rock who had released their last record forty years earlier, remains instructive. Their preferred interventions may differ somewhat, but Harris and Grainge both make appeals to preserve a system through which exceptional aesthetic and commercial worth are rendered mutually coextensive, enabling trailblazing artists to convene with their avid supporters at scale. Collapsing the artist-fan binary and supporting amateurs to pursue the practice of music for its own sake, regardless of what anyone else thinks of the outcome (as in, for example, radical versions of 'cultural democracy'), is categorically *not* their business: to do so would be 'impoverished'.

Importantly, this was not a mere war of words and principles; the prospect of impoverishment was a very real one. The material battleground was the UK Digital Economy Act 2010, which would grant statutory powers to locate specific instances of copyright infringement, enabling the possibility of disconnecting and prosecuting the most persistent individuals. Recording industry executives and other rightsholders lobbied hard for this change, with reports of private meetings between senior executives and ministers. But it was acknowledged that a more delicate approach was required, with a graduated response of sanctions preceded by 'educational' warning notices. It was not just Harris who expressed concerns. In one trade body meeting I attended in 2015, it was clear that sensitivities remained high within some industry corners. Recalling that it was not so long ago that public opinion of the industry 'made you want to hang yourself', board members worried that the proposals would 'alienate consumers' at a time when 'the market's just started to like us again'. But the very fact that the opinions of 'the market' mattered at all was new: no such legitimacy had been necessary when the only means of accessing music was the radio or the record store. But as digital tools began making the means of reproduction and distribution widely available outside of this bottleneck, the dominant actors had lost control.

At stake in this shift, then, was the communicative challenge of asserting the public value of copyright as a means of aligning protections for creators with fans' passion for music. Symptomatically, I suggest, this converging imaginary has been echoed by those paying attention to industry within much of the academic discourse associated with popular music studies (PMS). From its mid-century beginnings, this sociologically inclined critical literature sought to account for the key links in the popular music value chain, understood to be songwriting, performance and arrangement; sound engineering and production; talent management, scouts and artists and repertoire (A&R); and promotion, music press, plugging and radio play (Hall and Whannel 1964; Frith 1981; Hirsch 1972; Negus 1992; Peterson and Berger 1971; Stratton 1982a; 1982b). Additions to the scholarship likewise mapped material changes in industrial structure over the intervening period. Exemplary here is the changing position and practice of recording studios (Hennion 1989; Leyshon 2014; Watson 2015), record shops (Du Gay and Negus 1994; Hracs and Jansson 2020) and concert promoters (Cloonan 2012). The increasing centrality of promotional culture (Meier 2016; Powers 2010) is evidenced by studies of taste-making cultural intermediaries employed by record company market research and promotion functions (Negus 1999; Powers 2012), as well as the heightened importance of music licencing and supervision in coupling music with brands, advertising, television and films (Klein 2008; Taylor 2012). Work on journalism and criticism (Brennan 2017; Forde 2001; Powers 2013), on collective management organizations (Street et al. 2018), and on trade bodies and policymakers (Bennett et al. 1993; Cloonan 2007; Harker 1997) has also broadened the scope.

Despite this expanding cast of characters, the experience of the musical *creator* (writer, performer and sometimes studio engineer) remains privileged by PMS, insofar as it is their presumed position at one pole of an irreducible art-commerce dichotomy that justifies scholarly attention. The 'economic' or 'industrial' side of the picture only appears to the extent that it supports creative artists or, more likely, places them under corrosive pressures (cf. Harker 1980; Toynbee 2000). This erasure of the economic speaks not only to a sense of professional identity but also to a particular set of ethical or political commitments, ones that track across scholarly and everyday accounts of music's significance to social life (Klein et al. 2017; Whelan 2014). As such, 'art-commerce' and 'indie-mainstream' binaries work to code actors and their actions implicitly (sometimes explicitly) as *good* or *bad* – or rather, worthy or unworthy of attention – only insofar as they are thought to articulate or impede the conditions in which typically vague and undefined notions of unfettered 'creativity' and 'artistic freedom' might thrive. This allows PMS' critical disposition to come into uncomfortable alignment with the self-image of powerful industry actors (cf. Wheeldon 2014), building coalitions among otherwise opposing positions. The stated commitments of the CEO, the journalist and the PMS academic all converge, I want to suggest, on a shared 'moral economy' (Banks 2006; Sayer 2000), settling on the promise to sustain music as a higher good.

Why should this matter? Is it not mere common sense? We can argue about how genuine such commitments might be but, ultimately, who would not want to support music and protect creativity from external pressures and constraints? These kinds of questions are central to the notion of 'moral economy'. By this, I refer not to a somehow 'more ethical' approach to a music economy – perhaps one driven by altruism or corporate social responsibility – but simply one populated by real people with passions and concerns, to whom 'music matters' (Hesmondhalgh 2013). Without them, there is no music economy to speak of. Modern economies are hugely complex, often inhuman, systems. Often, their inhumanity is derided as the malevolent work of a manipulative cabal (whether corporate greed or some more sinister conspiracy), including by some musicians. Sometimes, by contrast, this systemic inhumanity is celebrated: the economy as a vast global information-processing machine, where the abstraction of decision-making away from any personal dimensions is seen as a sophisticated achievement. Either way, 'economy' is typically viewed as an entirely separate realm from the fleshly stuff of music and its lifeworld.

From a moral economy perspective, this view is unhelpful. True, the rise of neoliberal governance over the course of the twentieth century was underpinned by a thin textbook model of human action as a calculation of maximal utility (Davies 2017; Milonakis and Fine 2009). As a market of markets, international finance was thus argued to regulate global capitalism better than the mercurial judgements of any single politician (Roscoe 2015) – at least until the dramatic collapse of 2007–2008 showed just how untenable this imaginary was (Tooze 2019). Popular music has been, and continues to be, thoroughly imbricated in all these dynamics, in ways that have yet to be fully articulated – and I do not attempt to articulate them here. For now, I want simply to signal that *all* economic activity – getting a job, purchasing a streaming subscription, buying a house, starting a record label, enforcing intellectual property rights, investing in derivatives – is charged with morals and norms (positive or negative) that make claims (implicit or explicit) about what is *legitimate* or *illegitimate*. Here I am placing particular emphasis on how such claims take shape in the context of *work*. Devoid of such dimensions, the 'system' would simply not run. That is, *value* is always embedded in *values*. Any attempt to change the system, then, had best be clear about what they are.

Working for the man

Much of this book is dedicated to examining the relationship between industrial reorganization and working life within popular music in the early twenty-first century. But the broader import of doing so is to explore how such changes intersect with what the sociologists Luc Boltanski and Ève Chiapello (2005) characterize as a 'new spirit of capitalism'. The phrase, borrowed, of course, from Max Weber's diagnosis of the Protestant work ethic, indicates a major problem with capitalism: people must consent to reproduce it under what are, in many respects, absurd conditions not of their own choosing. Without ethical justification for participating in the 'capitalist order', citizens of all stripes are liable 'to find their everyday universe uninhabitable' – 'spirit' is that which hence 'helps to justify this order and, by legitimating them, to sustain the forms of action and predispositions compatible with it' (Boltanski and Chiapello 2005: 10). For Weber (2001: 17), by the end of the 1800s, the 'spirit of modern capitalism' describes a situation where the accrual of capital was no mere matter of personal gain but instead 'takes on the character of an ethically coloured maxim for the conduct of life', a contribution to a higher virtue. Small-scale bourgeois inventor entrepreneurs like Benjamin Franklin, committed to the broad concerns of economic accumulation through progressive technological advance and a disciplined dedication to one's work, were idealized and generalized. Equally, following the economic historian Albert Hirschman (1997), Boltanski and Chiapello note how the pursuit of profit was increasingly invoked by liberal political philosophers and statesmen as a civilizing technique, taming the violent passions that led to war and instead aligning populations in a shared 'common good'. As such, they remark, 'the emergence of capitalism presupposed the establishment of a new moral relationship between human beings and their work [...] in the manner of a vocation, so that, regardless of its intrinsic interest and qualities, people could devote themselves to it firmly and steadily' (Boltanski and Chiapello 2005: 9). It required, that is, a sense of moral 'spirit', in which one's everyday conduct, in life as in work, could be allied with a broader sense of the common good.

What does all this have to do with the industry of music? Am I suggesting that pop exists merely to furnish capitalist exploitation with ideological cover? This narrative – a caricature often crudely ascribed to the Frankfurt School critique of the 'culture industry' as a form of 'mass deception' (Adorno and Horkheimer 1997) – retains some popular currency after all, if only as an insult. But my aim here is quite the opposite. Rather than dismissing commercial music, I want to take it seriously and describe how, as a complex production system, it is hardly innocent of these broader dynamics of socio-economic transformation. Yet more than this, I stress that it has its own logics and justifications, mediated by musical passions and commitments – its own ethical 'spirit' – which formed a key site of struggle and renewal during the period covered by this book. To explore this, it is worth setting out further historical context.

In Boltanski and Chiapello's diagnosis, a change in spirit occurred with the development from the 1930s of what is often termed 'Fordism', centred on large industrial firms and mass production of standardized goods, alongside the growth of white-collar work in the form of the management profession (cf. Beynon 1993; Braverman 1998). The Fordist arrangement responded not only to developing technologies and a need to marshal increased productive capacities but also to criticisms of the brutality of this form of mechanized labour exploitation, alongside the rampant inequalities it enabled, of the kind to which the likes of Marx and Engels gave voice. Greater protections and 'a fair day's work for a fair day's pay', were fought for through the trade union movement that had been buoyed by concentrations of workers in mills and factories. For those employed in such contexts, the long hours of hard, repetitive work within a highly stratified division of labour were offset by the opportunity to improve their standard of living with relative job security, reductions to the working week and, importantly, improved pay and spending power: the capacity to buy the goods they were producing. Ideal Fordist organizations 'constituted protective environments not only offering career prospects but also taking care of everyday life', while represented by 'an ideal of industrial order embodied by engineers', supplemented by a sense of civic duty: 'institutional solidarity, the socialization of production, distribution and consumption, and collaboration between large firms and the state in pursuit of social justice' (Boltanski and Chiapello 2005: 18). Artistic craft was also folded into this complex division of rationalized labour, propping up highly processed and standardized forms of both cultural production and cultural consumption (Adorno and Horkheimer 1997). Nonetheless, the sheer productivity of such organizational forms, and the desirability of the goods they produced, were enough to counter emergent critiques. The early record companies were taken as one significant expression here, whether in the routinized songwriting factory of the Brill Building; or, later, in the Hitsville studios of Detroit's Motown, which founder Berry Gordy famously claimed to have modelled on Henry Ford's production line. But the ideal of job protection and broader social purpose, undergirding continuities between work and leisure, and between firm and state, equally carries through to the contemporary music corporation.

The experience, and subsequent critique, of this moral order is captured in Roy Orbison's 1967 song 'Working for the Man'. The Man in question is the authoritarian boss in a factory where work is menial and repetitive, and employees are kept strictly to time with few breaks. He also happens to be the father of the protagonist's love interest, who secretly keeps him refreshed, and with whom he elopes at the end of the working day. More figuratively, in the popular vernacular of the time, The Man signifies a rising resentment towards this standardized, often punitive form of production and its attendant social order. While it is possible to read the song in strict Marxist terms – as an expression of material antagonisms between labour and capital – the phrase is malleable enough to be applied to any authority figure in a bogus natural hierarchy: the state, the police, the corporation, the husband, the banks, God, the historic enslaver, the global 'system' or simply the social conformity of an older generation. In other words, The Man resonates because it embodies an abstract and assumed source of oppression, rendering it available as an object of critique for a growing counterculture.

Against the conformist system, an ethos of social transgression, plural identities, and anti-establishment resistance has resonated through popular music mythology (Klein et al. 2017; Doggett 2007). But, if read against one diagnosis of this era, widely circulated among countercultural radicals – the psychoanalytic reading of social relations produced by Adorno's compatriot Herbert Marcuse (1991) – 'Working for the Man' remains conservative. After all, the post-war settlement of secure work was predicated on an idea of fixed gender roles and the nuclear family that the song hardly unsettles. Orbison's factory worker is alienated, bored and sexually frustrated – but he is no radical. The punchline of the final verse sees the downtrodden employee assert ownership over the boss, and ultimately the company, by claiming the daughter as his own ('they're both gonna be all mine'). And so the song hitches its worker's critique of the conditions of industrial production onto the tacit reproduction of patriarchal structures and property relations that inevitably flow through them in its elated closing lines: 'Yeah, I'm gonna be the man!' Through Marcuse's Freudian lens, this fantasy enables him (and the listener) to reconcile the excesses of the Pleasure Principle with the paternal authoritarianism of what he calls the 'Reality Principle', internalizing the constraining 'cultural superego' of the father figure: the cop inside your mind. Marcuse's more radical prescription was to unleash the pleasure: to follow one's desires more fully rather than adapt and sublimate them to capitalist production. Capitalism, however, had other ideas.

The artists' critique

According to Boltanski and Chiapello (2005: 27), when critiques proliferate, they allow the capitalist order 'to find the moral supports it lacks and to incorporate mechanisms of justice whose relevance it would otherwise have no reason to acknowledge'; hence, by recognizing and incorporating certain critical elements, capitalism regenerates its 'capacity for survival'. The Fordist compact of the 1930s had been informed by social critiques of inequity, generating demands for greater democratic representation, social justice and freedom from coercive repression. But by the 1970s, the internal contradictions of this arrangement were becoming increasingly visible, with rising inflation, widespread industrial action and unemployment associated with energy crises and the flight of manufacturing. This prompted a wave of socio-economic restructuring that was, in many respects, brutal: dramatically shrinking welfare supports and explicitly seeking to break workers' bargaining power, especially through trade unions.

Such reforms found uneasy support from the countercultural movements of the late 1960s, which had generated what these authors call an 'artistic critique' of societal alienation, rooted in romanticism and bohemianism, that sought to recognize and empower individual autonomy and freedom – opposed to diagnoses of class struggle, in which class could appear as crudely undifferentiated. By contrast, demands for greater understanding of demographic differences, and more outlets in which to freely express libidinal desires, led to expanding business opportunities (Frank 1997). Popular music was an important vehicle for these ideals, such that some were able to draw on their aesthetic expertise to find employment. Record companies began hiring representatives from the counterculture, for example, such as 'company freaks', charged with translating between production and consumption (Powers 2012). Charged with keeping the company in touch with the sounds of the street, both channelling and influencing emerging cultural habits, these 'new cultural intermediaries' were capitalism's 'ethical *avant garde*' promoting a 'morality of pleasure as a duty' (Bourdieu 1984: 367). In such ways could the Beatles' 'Revolution' travel from countercultural anthem through to Nike advert soundbed – even if this was a shorter distance than might be popularly imagined (Bradshaw and Scott 2018).

Political theorist Nancy Fraser (2013) argues that the so-called 'second wave' of feminism – which sought to break with the normative family unit underpinning men's (unionized) labour while domestic work went unrecognized – was a significant thread here, increasing opportunities for women to participate in the workforce. As such, numbers of women in music were also growing, if predominantly in secretarial and publicity roles (Leonard 2016; Parsons 1988; Steward and Garrett 1984). But the gendered tensions of such changes were complex and classed, perhaps best encapsulated by some rather less creative workers further down the value chain. In 1976, a dispute over equal pay between the Transport and General Workers' Union (TGWU) and the EMI Records packing and distribution centre in Hayes, Middlesex, took on a different hue after the Sex Pistols' notoriously obscene television interview with Bill Grundy. Sixteen female TGWU members 'refused to handle any material' by the band, on the grounds that it was 'degrading' – two women eventually negotiating an exemption from doing so – and thereby 'using industrial militancy not only to gain wage equality' but showing 'they were also willing to withhold their labour in order to preserve a particular notion of femininity and working-class respectability' (Gildart 2013: 134). Such instances demonstrate that, if the broad direction of travel was clear, any transition from one era to the next is not quite so simple.

Economic stagnation and clashes of values increasingly caused the efficiencies and long-term contracts of rational bureaucracies to be questioned. In response, appeals to the more dynamic 'flexibility' of 'projects' and 'networks' grew: keywords for what Boltanski and Chiapello consider to be an emerging 'new spirit' over the 1980s and 1990s. Although primarily signalling a set of pressures on managers and civil servants to cater to consumer demand (Du Gay 2007), this spirit nonetheless gelled well with broader appeals to creativity. Hence, corporate managers might look to countercultural icons like the Grateful Dead for authentic market engagement ideas (Bowen and Siehl 1992), while musicians and independent labels modelled new modes of work and organization (Hesmondhalgh 1996; Stahl and Meier 2012), paving the way for the coming (suggestively-named) 'gig economy'. In the United States, such appeals were crystallized by the urban theorist and consultant Richard Florida (2002). He diagnosed the displacement of the traditional working classes with an ascendant 'creative class' of cultural, media and knowledge workers: a *problem-defining* 'super-creative core' of artists, neo-bohemians and technical innovators, generating new ideas and intellectual property, supported by an outer ring of *problem-solving* 'creative professions', from lawyers to business consultants, applying established knowledge to new situations. In Florida's telling, the creative class took from the earlier hippy and punk values as meritocratic DIY neo-bohemians, gathering to assemble projects (often in California) with like-minded others who prized 'talent, technology and tolerance'.

In the UK, this shift went under the banner of the 'creative industries', a policy programme inaugurated by the New Labour government in 1998, drawing from earlier work on localized cultural industries (Banks and O'Connor 2017). According to one report published by the Labour-associated Demos think tank, for example, the 1990s had been the era of 'the independents': freelance cultural entrepreneurs and micro-enterprises setting their own goals, not being organizational wage slaves (Leadbeater and Oakley 1999). Both agendas sought to address the challenge of urban deindustrialization since the 1970s, with an increasingly globalized economy driving the relocation or outsourcing of physical manufacturing and distribution processes to cheaper locations, especially in the Global South. Attention shifted instead to the apparently 'post-industrial' activities of design, human-centred services and culturalized experiences. Not only were these occupations considered a 'high value' source of global competitive advantage, they were also thought to offer more fulfilling and self-realizing forms of 'good work'.

Attention to the working *conditions* behind this buzz only came onto the agenda rather belatedly. By the end of the 2000s, a recognizable 'turn to cultural work' was underway, whereby, after having long been sidelined in both media and cultural studies and in the sociology of work and organization, 'the labouring lives of people working in the cultural and creative industries are now firmly on the research agenda' (Banks, Gill and Taylor 2013: 2). Such research typically emphasized the precarity, low or irregular pay and the absence of benefits or social security that characterized much of this work – and the inequalities it thereby produced and reinforced (Gill and Pratt 2008; Eikhof and Warhurst 2013). From a governmental perspective, according to McRobbie (2016: 35), this notion of 'independence' indicated a workforce 'being trained up to pave the way for a new post-welfare era'. In such ways, by the start of the 2010s,

> for good or ill, workers in media, design and the arts are routinely held to exemplify the 'worker of the future'. [. . .] Artists and 'creatives' more broadly are said to embody the new form of constantly labouring subjectivity required for contemporary capitalism, in which the requirements for people fully to embrace risk, entrepreneurialism and to adopt a 'sacrificial ethos' are often linked to an artistic or creative vocation. (Banks, Gill and Taylor 2013: 3)

Musicians and their representatives were, among others, understood to be at the forefront of the new spirit of capitalism as exemplars of the good life under post-Fordism. Although not always compliant, they represented a more flexible labour force for a more flexible business model: committed to the (sometimes transgressive) pursuit of aesthetic novelty, rendered as product differentiation, under an increasingly neoliberal approach to governance driven by spreading market competition (de Peuter 2014).

Uncreative music

> One of the things that's coming out [of the consumer studies] at the moment is the notion of work. We've got a country that seems to be talking a lot about work and working hard. That's a really interesting bit of cultural discourse that I think probably has its roots in coming out of recession. Politics is never far from it – not party politics but ideas of young people making their own destinies, young people not trusting the state, do it themselves. I was talking to someone about Iggy Azalea, 'Work' – you know, [rapping] 'work, work, workin' on my shit . . . ' – I think

it's gonna be a really interesting area to explore. And actually, as much as it can sound like intellectual masturbation, it's actually really relevant. Because we're in the business of creating culture. And what we do matters.

Jeff, interview (2014)

Placing the conditions of music's production within a broader historical context is an interesting academic exercise. But it is also something that concerns the majors themselves. Consumer insights manager Jeff, one of the *new* 'new cultural intermediaries' working for the contemporary corporate Man, is paid to inform marketers about things that 'matter' to people. The difficulty of crafting one's destiny through hard work is one such thing that matters, something he wants to connect to the very recent financial crisis and the subsequent austerity conditions imposed by governments in response. This is not simply then, in his overtly libidinal metaphor, 'intellectual masturbation'. What he does 'matters' because, he implies, culture steps in where states have failed to inspire trust. Crucially, it is the very 'notion of work' that Jeff considers 'an interesting area to explore'. The problem is not simply the difficulty of finding a job, the *concept of work itself* is under pressure. He finds one plausible answer in the recording artist – who is not an employee but a contracted freelancer and business entity in their own right. Perhaps an aesthetics of 'resilience' (James 2015), with its mantra of 'working on my shit', offered by major label artists like Iggy Azalea – or Britney Spears, or David Guetta, or Wiz Khalifa, who all released work-themed anthems in this period – offers something else to believe in (Campagna 2013).

The problem of work also came to the attention of PMS in roughly the same time period. Beyond the 'art' of making music, authors have discussed legal and economic issues of career sustainability, training, working conditions, remuneration, disputes, protections, representation and so on. They have raised visibility, for example, of how creative freedom is restricted and negotiated in changing terms of recording contracts (Arditi 2020; Marshall 2013b; Stahl 2013; Tschmuck 2009), negotiations of online fan relations (Baym 2010; Morris 2014) and especially the pressure for artists to develop entrepreneurial strategies and intermediary skills (Cluley 2011; Haynes and Marshall 2018; Hracs 2015; Kribs 2017; Morgan and Wood 2014). This has echoed real changes in the status and representation of popular musicians themselves as workers, rather than 'mere' artists or entertainers (Frith 2017). The UK Musicians' Union (MU), for example, which ostensibly represents the rights of recording artists, songwriters, performers and 'session' musicians alike, has echoed other trade unions with a declining membership over the twentieth century. The MU has struggled to respond to technological change, particularly the diversified income streams that are increasingly reliant on the collection of royalty payments rather than the fixed salaries of orchestras or the piecework of session musicians. This has led to territorial disputes over who or what 'counts' as a musician and a shift 'from being largely a workers' organization towards becoming a music industries' one' (Williamson and Cloonan 2016). More recently, public and policy discussions over the rates of remuneration generated by streaming have also brought musicians' economic conditions to the fore (Hesmondhalgh 2021). A major fault line remains, nonetheless, between social critiques that stress equity and distributive justice within the employment relation, and an alternative version of 'work' filtered through the lens of protecting the creation and long-term ownership of desirable copyrights. That is, between those whose existence relies on waged work and those that depend on income generated through intellectual property rents.

It should be instructive, therefore, to explore this tension among employees of a major music rightsholder. A handful of studies have taken those working at the less glamorous end of the spectrum as objects of interest in themselves: business affairs executives and warehouse operatives (Negus 2002), human resources (HR) departments (Colbourne 2011) and label interns (Frenette 2013). Such work has only risen in significance. For instance, Jeremy Tunstall's (2000: 22) overview of media industry occupations noted that, while 'media work is now more complex, more specialized and faster-changing [. . .] in other respects all media work has become more similar and also more similar to office work in general'. Hesmondhalgh (2007: 205) recognized how 'relatively humdrum (though often essential) work in the cultural industries has increased, partly because the cultural industries are gradually becoming a more important part of economies [. . .] but also because of the continuing importance of manufacture and wholesaling'. Yet critical accounts have struggled to do justice to such work by incorporating it within existing conceptual frameworks – where they even recognize it at all. Some are prone to equating this complexifying division of labour simply with a 'ballooning [. . .] administrative sector' and yet further evidence of capitalism's proliferation of pointless 'bullshit jobs' (Graeber 2013).

More charitably, Mark Banks (2007: 190n) observes – although tellingly, in a footnote – 'the necessity of undertaking research into "non-creative" workers in cultural industries': the many 'manual, clerical, administrative, technical and managerial staff making vital contributions to the production of recordings, programmes, movies, paintings, garments and the like'. However, while also 'subjected to various degrees of freedom and constraint' and 'differential rewards and recognition according to their real or perceived contributions to the production process', he judges it 'likely that these non-creative workers are positioned within a more conventional (and well-researched) division of labour, with their work organized around a series of standardized, routinized and mundane tasks', where 'the art-commerce relation is not so pronounced' (Banks 2007: 190n). The dominant approach thus remains to position 'creative personnel' as the 'primary workers in the production of symbols, information, entertainment and meaning – the core products of the creative industries'; consequently, we still know 'little about the conditions of such [non-creative] workers and this could be a significant topic for future research, if theorised well' (Hesmondhalgh 2013b: 259).

In response, we might return to Howard Becker's classic sociological formulation of the 'art world', an influential means of destabilizing the primacy of the lone artist-genius. Cultural production, he writes, is defined by 'cooperative activity, organized via [. . .] joint knowledge of conventional means of doing things', with a key issue being 'to decide which of all these people is the artist, while the others are only support personnel' (Becker 1982: x). The latter activities 'vary with the medium: sweeping up the stage and bringing the coffee, stretching and priming canvases and framing the finished paintings, copy editing and proofreading'; or equally, 'technical activities – manipulating the machinery people use in executing the work – as well as those which merely free executants from normal household chores'. This founding cut is thereby a negative one, rendering 'support' as simply 'not creative': 'a residual category, designed to hold whatever the other categories do not make an easy place for' (Becker 1982: 4). Becker's insight is that this cut is not necessarily imposed from outside, but labelled from within, performed through the routine conventions of the art world. Vicki Mayer's (2011) study of 'invisible labour' in television industries began from this internal separation, as institutionalized in standard Hollywood accounting procedures. The latter distinguishes between the fixed costs incurred by employing 'talent' (creative actors, screenwriters, producers) 'above the line' and the variable costs of craft or technical workers and others 'below the line'. These are useful categories insofar as they underscore the distinction (the 'line' on the accounts statement) between those who assign control of their Intellectual Property in exchange for 'residuals' (or a one-off fee) and those whose labour power is exchanged for a wage. For Mayer – whose exemplars range from television set assembly workers to talent scouts and regulatory committee volunteers – the experiences of below-the-line workers, as well as the value that is extracted from them, have been unjustly neglected.

Authors working in a critical relation to Pierre Bourdieu, meanwhile, emphasized processes of cultural *intermediation* (cf. Smith Maguire and Matthews 2012), a term that might help sensitize us to the contexts within which normative divisions between creative and support work are established. In an important intervention, Keith Negus (2002) uses the examples of business affairs and warehousing in music, each representing 'non-creative' staff at different points in the production chain. Because they are not aesthetic decision-makers, he argues, both are kept at a distance by scholars and cultural workers alike – yet both demand attention for the crucial roles they play, respectively, in the signing of acts and the manufacture and distribution of cultural goods. Anne Cronin (2004: 351–2) similarly challenged a 'disproportionately intense focus on Creatives in the advertising literature', which 'does not adequately address the significance of Account Managers, Account Planners or Media Buyers within the industry'. Such roles attune us to the materialization and stabilization of ideas, beyond their initial generation, through what she describes as multiple hybrid commercial-cultural 'regimes of mediation'.

From a different perspective, Chris Bilton (2015: 154) offers a valuable reflection on the 'creative thinking' advocacy of popular management texts, which legitimizes 'a division between a creative minority and an uncreative mass': that is, 'individuals, institutions or procedures outside the process of idea generation, including intermediaries, administrators and "bureaucracy"' who become various 'blockages', associated with 'self-doubt, resistance to change, value judgements and inhibitions'. Bilton (2015: 163) poses a more dialectical notion of 'uncreativity' – variously found in 'resistance, criticism and negative thinking'; 'the everyday grey of normal office life'; 'privacy, containment, routine and predictability' – as neglected but constitutive moments in innovation processes. Taking this forward, then, rather than treating support work as a 'necessary evil', more positively, we might explore non-creative tasks *in their very uncreativity*. Whereas popular discourses of 'innovation' champion *disruption*, they obscure the wealth of vital *maintenance* that goes on behind the scenes to keep things running (Russell and Vinsel 2016). Recognizing the work of maintainers, then, aligns with an ethic of care: regularly attending to those instances of rust, dust, cracks and corruption that litter our world, through which life gradually becomes unliveable (Mattern 2018).

Perhaps more so than other cultural industries, popular music histories remain tied to heroic narratives of artisans and record men: the creative disruptors. But while the complex division of screen-based labour is protected, negotiated and visibilized in collective bargaining and end-credits, there is little such recognition of collaborative production systems in popular music beyond the roll-calls and namechecks of album liner notes or awards speeches or the invisible (and fiercely contested) field of metadata. With honourable exceptions, we still know strikingly little about how a range of vital elements – databases, financial calculations, rights administration, warehouse distribution, promotional upkeep, product assembly, event maintenance, risk calculations, enterprise training, trade legislation, duty of care and so on – have played out across, and been mediated by, the practices, sites, production systems and spirits of various 'musical capitalisms' (Born 2013), old, new and yet-to-come. Finding out might lead us into more infrastructural and logistical dimensions of particular materials and resources, processes of assembly and circulation (and, of course, breakdown), which form the very conditions for any music to emerge at all (Sterne 2014; Devine and Boudreault-Fournier 2021).

But we need not even go that far. Taking care of music as a viable collective enterprise is more simply about maintenance. It just so happens that at least some of this caretaking happens within and around transnational corporations. Much of this activity is unglamorous, restrictive or simply boring – perhaps to the extent that we might question whether it is really 'cultural work' at all. Crucially though, as Mayer recognizes, questions of professional identity (whether individuals identify in some sense as belonging to a 'music industry') act as a central resource for value creation. They too perform the cut of judgement described by Becker (1976: 226–708), as 'integrated professionals': those who 'do not violate convention sufficiently to disrupt coordinated actions' but nonetheless 'fit easily into all the standard activities that the [art] world carries on' precisely *because* they 'habitually use the conventions on which their world runs'. Unravelling the 'complex mirroring of producers with political economy, and of identity with labor value' (Mayer 2011: 1) may take us beyond simply reinstating unheard voices towards a wholesale reimagining of how participation within cultural production is understood, accounted for and justly rewarded. To ask what makes music's maintenance work *meaningful* is also to ask how music's meanings are maintained.

The lost spirit?

The world economy was brought down by derivatives and bad debt; music has been depleted of meaning through derivativeness and indebtedness. The imbalance in Western economies towards financial and real-estate speculation meant that too much of the wealth generated was meta-money [. . .]. Similarly, the profusion of hyper-referential bands and micro-genres whose stylistic involutions are understood only by hipsterati and bloggerati resembles the 'complex financial instruments' that only a handful of people in Wall Street and the City of London comprehend.

Speculation led to a spectralisation of the economy: money that has only the most tenuous, remote relation to the material world. [. . .] Culture, as the superstructure to the economy's base, reflects the gaseous quality of our existence. The insubstantiality of the economy revealed itself, horribly, a few years ago. We are still waiting for the music-about-music bubble to burst. (Reynolds 2011: 492–3)

Popular music has been an important system for both contesting and renewing the spirit of capitalism, one that continues to be dominated by corporate oligopoly. Renovating this system's own spirit, however, was experienced as a significant challenge for the first time in the early years of the twenty-first century. The corporate behemoths have not retreated; if anything, they are yet more central to contemporary popular music's economic life – its consumption, production and, perhaps especially, its regulation and legitimation. Increasingly entwined with global 'asset manager capitalism' (Braun 2021), present-day battles concern carving out space in the landscape dominated by technology companies and financial investment institutions like Blackrock and KKR. The period during which their profits tumbled, amid what Harris called 'the great chaotic utopia' of emerging digitization, was, in retrospect, relatively short-lived.

Ironically, this period coincided with diagnoses from cultural critics – no friends of corporate music – that popular music's most exciting and innovative tendencies had simply run out of steam, replaced with endless recycling and nostalgia-fuelled 'retromania' of cover songs, reissues and anniversary tours (Reynolds 2011). Such criticism maps against what some saw as a decline of 'narrative', and corresponding rise of the reflexive cultural form of the 'database' (Manovich 1999) or 'list' (Young 2017) – or indeed 'playlist' (Rekret 2019). 'While 20th-century experimental culture was seized by a recombinatorial delirium, which made it feel as if newness was infinitely available', wrote Mark Fisher (2014: 8), 'the 21st century is oppressed by a crushing sense of finitude and exhaustion', haunted by ghosts of those earlier potential futures. He coupled this fate to declining emotional and financial investment in culture and to a wider mental health crisis – echoed among working musicians themselves (Gross and Musgrave 2020). If cultural industries' capacity to offer innovative narratives of alternative futures was depleting, the 'spirit of capitalism' was in danger not of renewal but of being 'lost' entirely – with potentially disastrous results (Stiegler 2014).

One might, in this context, point to constraints placed on expenditure or on the freedom of record men to develop unknown talent, as part of corporate risk mitigation strategies – much as the likes of Grainge and the BPI would protest.[1] Perhaps. But a more convincing explanation would trace broader structural shifts, tied to the spread of secondary rights exploitation opportunities: proliferating 'synch' licencing (music for TV, film, advertising, video games and so on); extended technical capacities and legislative powers to capture and police background music (in retail settings, for example); and then, of course, streaming. If platforms like Spotify or Tidal were to have any chance of success, attracting both users and investors, this would be predicated on the ability to sign up the catalogues of rightsholders – including the three major labels, in exchange for significant equity stakes (Vonderau 2019). This, in turn, enabled new functions, such as the monitoring of listening behaviour at scale and the formation of music recordings as a new asset class for further financial speculation (Nicolau and Wiggins 2022). The initial public offerings of companies like Universal and Spotify take the confrontation between 'music' and 'tech' into new political-economic terrain – again entirely enabled by their capacity to control ownership, manage usage and provide access to vast catalogues of rights (Watson and Leyshon 2022).

Yet none of this could take place without work: the mundane, administrative, uncreative work that sits alongside artistry, executive decision-making and cultural intermediation. In this book, then, I am drawing attention to major record labels not just as exploiters of artists or market monopolists but as employers in their own right. Who works for them, I ask, in what capacity – and, crucially, why? How is it that members of the so-called creative class would seek employment performing routine tasks in large bureaucratic corporations? What does that maintenance work look like on a day-to-day basis? How do they make sense of all this, especially amid times of industrial and economic crisis? In what ways does music matter to them? What justifications and critiques do they launch as a result?

The remainder of the book sets about answering these questions.

2

Access all areas

The internet revolution

We're living in the internet revolution, aren't we. We're literally in the heart of it.
[. . .] It's kind of recognised that we're gonna be living in an access-based world,
rather than an ownership-based world, right? [. . .] We know that. Streaming's king
for music and TV . . . well, the kids are buying vinyl again . . . You know, there's
been a gradual increase in vinyl sales, which does hint that we have a collector side
to our nature as human beings. Some of us do anyway. [. . . But] I think we know
that's gonna be the way forward [. . .] whether you bundle in your music package
within your TV/broadband line, we know that's the way it's going, pretty much. But
for the general public, it's still bedding in, it's still multiple.

Ben, interview (2014)

I was introduced to Ben, an artist manager for a major-signed band, by a mutual
contact backstage at a festival and trade conference for artists, decision-makers
and thought leaders in music. In an interview a little later, he expressed to me a
sentiment that captured the era. The 'internet revolution' was well underway, and
he was 'in the heart of it'. The world of ownership was on its way out. Access was
'the way forward', whether we realize it or not, and whatever form it takes, this
is 'the way it's going'. The future had arrived. He was hardly the first to make the
observation. For some time, it had become seemingly impossible for academic
or popular writers to refer to the record industry without adding a prefix. It
was either *new* (Hughes et al. 2016; Marshall 2013b; Taylor 2013; Winter 2012),
digital (BPI 2013a; Hesmondhalgh 2009; Hracs 2015; IFPI 2013; Kusek and
Leonhard 2005; Meier 2017; Stahl 2013) or both. Accordingly, this was widely
agreed to be a period of *transformation* (Leyshon 2014), *disruption* (Nordgård
2018; Wikström and DeFilippi 2016), *crisis* (Gilbert 2012; F. Goodman 2010)
and *revolution* (Sun 2019). Sometimes this took the form of a 'rise and fall' or
'life and death' narrative (Barfe 2006; Rogers 2013; Southall 2009); elsewhere

it simply alluded to a 'shift' or 'change' in structural balances (Benner and Waldfogel 2016; Mulligan 2014; Negus 2015). In one form or another, the music economy required some serious *rethinking* (Baym 2010; Morrow et al. 2022; Williamson and Cloonan 2007).

Among these various accounts, perhaps the most important keyword was *access*. Primarily, this referred to a change in the central modes of consumption, production and accumulation – that is, the methods through which value is located, created, extracted and circulated – or, in more common industry parlance, the 'business model'. Reclining on his fifth-floor office sofa, corporate commercial partnerships manager Nick identified 'growing access models' as 'a strategic focus for us' – quite simply because this is 'the way the world is now':

> *Because it's moving towards access, a marketing opportunity and a business opportunity is kind of the same thing. Because there's no sale, you know. . . . It's not: 'I've seen this ten times on YouTube so I'm gonna buy it'. It's 'I've seen it ten times on YouTube'. That's it. That's the business.*
>
> Nick, interview (2015).

Analyst Mark Mulligan (2011), meanwhile, had 'no doubt that Access based models are the future of music' – although, like Ben, he hedged his bets with the caveat that 'Ownership based models will continue to play a pivotal role'. The record industry trade bodies agreed (BPI 2013a; IFPI 2013). These references to 'access', then, indexed changes in consumption preferences and listener behaviour. As Nick put it, 'there's no sale'. Rather than buying music objects – a record, cassette, CD or download of a digital file – in traditional retail transactions, people were increasingly happy to rent it, typically via subscription, on a time-limited but renewable basis. They simply preferred their engagement with music to be as smooth and frictionless as possible. A more fluid metaphor was often used. David Bowie's early 2000s comment 'that in the near future music was to become akin to a utility like water or electricity' had 'accurately predicted' changes in revenue generation, according to Andrew Leyshon (2014: 159). Agreeing with Bowie, Patrik Wikström (2012: 11) lamented that markets for utilities like water supply were characterized by 'fierce competition and low profit margins' and hence 'not a very attractive place to do business'. The technology platforms and major record labels wading into the flow economy of the 'stream' disagreed.

Digital disruption, then, aligned with a broader spirit of the time, during which 'access' achieved a discursive currency well beyond music and media, resonating with far broader appeals to participation, equality and transparency (Carpentier 2015). Long-standing walls would be broken down and clear boundaries separating interior from exterior worlds replaced with more permeable membranes of regulated flows. The critic and analyst Jeremy Rifkin (2000) diagnosed a coming 'age of access' – primarily in enterprise operations but increasingly across consumption habits, lifestyles and subjectivities – predicated on temporary 'paid-for experiences', in which culture was to be at the forefront. For the philosopher Gilles Deleuze (1992: 5), much earlier, social life was increasingly conditioned by 'codes that mark access to information, or reject it' – these 'passwords' would govern the movement of bodies through buildings, urban space and computing systems. More grandly, this logic would reconfigure the linear trajectories of biographical and institutional life itself to align with the corporation, which, he was horrified to learn, now had a 'soul'. Linear movements through school or university into employment were being replaced by the constant remaking of oneself through 'perpetual training' across multiple work settings – all provisional on having the right 'access codes'.

The purpose of this chapter, then, is to explore 'access' as a keyword for a changing industry from three perspectives. First, taking the position of a potential consumer or audience within the normative industry understanding of shifting business models. I suggest that the relationship of 'ownership' to 'access' is indeed important – but, even understood on its own terms, far more complex and iterative than is commonly taken for granted. Second, taking the view of the prospective employee, access to the major label workplace also changed, in line with broader cultural industries labour markets. Here too, change was not straightforward: while literature on the subject tends to highlight freelance creative entrepreneurship, both in veneration and lamentation, I focus on routes to traditional waged employment contracts. Finally, I take the perspective of the researcher, looking to access industry in order to study and generate knowledge about it. Again, the simple challenge of how to 'get in' is complicated by uncertainties over where 'in' actually is and by questions of how to interpret and make sense of that world once we do.

Equally though, knowledge is also sought by industry actors themselves. Professionals want explanations and discursive scripts that help them navigate shifting commercial cultures and make informed decisions – as Ben's hesitant appeal to lay theories of access and 'human nature' suggests. As such, his rhetoric is also instructive because, for all the confident revolutionary bluster, he remains uncertain. People are buying records again, it seems. With a quasi-anthropological air, he wonders about the resilience of 'the collector side' to human nature. Revising his teleology, he concedes that, even if the future is on its way, it is 'still multiple'. Rather than accepting or debunking the mainstream 'access' narrative, then, this chapter follows Ben's observation and multiplies it. My threefold approach scaffolds the three key themes of the second part of the book, furnishing us with the coveted Triple A 'Access All Areas' pass – the code or password – that we need to head backstage, behind the scenes, crossing the membrane and entering the world of the major label. We should hold on tightly to this pass, and the plural concerns it registers, as we move through the remainder of the book, stepping back and forth over multiple organizational thresholds. First, though, I want to turn back the clock to 1999: as apt a moment as one might hope to find in order to understand the millenarian zeal for record industry disruption.

Back to '99

As thresholds go, clearly, the transition from one century to another is an artificial one. In retrospect, however, the evidence for 1999 does seem to mount up. This year saw the launch of the peer-to-peer (P2P) file sharing service Napster, the usual culprit named for popularizing a rise in illegitimate MP3 filesharing. A correlated fifteen years of decline in revenues from sales of recordings was countered by a growth in income from synchronization, brand partnerships and other forms of rights exploitation – a tipping point captured by Moby's album *Play*, also released in 1999, famous for being the first to licence all its tracks to a TV show, film or advertisement (Klein 2008). Meanwhile, 1999's merger of Polygram and Universal Music Group collapsed the 'big six' into the 'big five' major record labels, which by 2013 had become the 'big three'. In the UK, the New Labour government had just completed its first creative industries mapping exercise and was getting to grips with policy initiatives geared towards intellectual property and creative work. And Keith Negus' (1999) *Music Genres, Corporate Cultures*, also published this year, presented a substantial study of the influence of music culture on firm strategy. Although it said little about the incoming technological reconfiguration, it nonetheless signalled a new hybrid cultural-economic approach in the field of popular music studies (PMS), alongside a new policy-facing agenda (Frith 2000).

The 'new year' is perhaps as good an example as any we have of a social construct: neither a 'natural law' of time nor an outright falsehood; rather, a human-made fiction that orders our organizational routines as well as our cultural and emotional lives. The passing of a millennium, with all its imaginative resonances, is even more so. For Prince, recording in 1982, this was, of course, the future – but an uncertain future. Seen from the vantage point of a Cold War moment in which 'everybody's got a bomb', 1999 looked like potential 'judgement day'. In 2018, by contrast, the millennium had become the subject of post-digital hyper-nostalgia for a simpler time, as in Charli XCX and Troye Sivan's '1999'. By then, 'Y2K' was a far more ambivalent 'aesthetic': an audiovisual jumble of pop culture, branding and consumer electronics to be mined for loose period signifiers. In 1999 itself, Will Smith (the artist formerly known as the Fresh Prince) was simply happy to ring in the coming 'Willennium' with a party and a cheeky nod to Prince Rogers Nelson over a sample from The Clash's 'Rock the Casbah' (also released, incidentally, in 1982). This was the high point of the 'End of History' diagnosed by political scientist Francis Fukuyama (1992) and, in pop, as in Jacques Derrida's (1994) 'hauntological' reflections on Fukuyama, 'time is out of joint': haunted by spectres of other moments and temporalities.

A charge of anticipation and uncertainty also manifested in the apocalyptic predictions of a coming 'millennium bug', or 'Y2K problem' – the programmers' concern that computers designed to hold only two-digit dates, rather than four, would be unable to cope with the year 2000 – with unknown and perhaps disastrous consequences. This seemingly minor glitch crystallized widespread *fin de siècle* anxieties towards an increasingly 'networked society' of home, business and government systems, and a global socio-economic order underpinned by complex and inscrutable technical protocols, intensifying government pleas to move information technology issues 'out of the back room and into the boardroom' (Knights et al. 2008: 301). All manners of private and public sector organization were thereby gripped by a lust for 'disruptive innovation': this described the argument that the most successful ideas do not cater to existing markets, they generate entirely new ones that overcome existing power-brokers and destroy what has gone before. Music was, of course, no exception.

1999, then, is a convenient point of entry for our inquiry – but not an easy origin story. It is just one possible moment in a passage of nonlinear time, powerfully overburdened by memories, aspirations, anxieties and, moreover, decisions already made, structures put in place, projections of risk and plans for managing them. This plural temporality runs counter to those intuitions about political, economic, social and technological revolutions as rapid and total, in a logic that is characterized by an almost religious zeal for transformation. For Fred Turner (2018: 162), writing about the tech imaginaries spread through MIT and the San Francisco Bay Area, the prophets of digital revolution were like 'seventeenth-century ministers', who 'instruct us that spiritual transformation remains the only way to find a place in a world we can't control'. According to the historian Jill Lepore (2014), notions of 'innovation' offer a simple modernist metanarrative, 'the idea of progress stripped of the aspirations of the Enlightenment' – but the related term *disruption* takes us into a more spiritual domain, 'holding out the hope of salvation against the very damnation it describes: disrupt and you will be saved' (Lepore 2014). Calls to disrupt, then, were meshed with a sense of the renewal of the spirit of (American) capitalism. They fit into 'a long tradition of millenarian prophecy in relation to new media', notes communications scholar James Curran (2010: 19): from facsimile newspapers and citizens' radio to camcorders, CD-Roms and experiments in digital television, apocalyptic predictions of decline proliferated; predictions that 'mostly American, proved to be wrong'. Beyond declinism, what makes such prophecies *millenarian* in tone is their eschatological sense of an inevitable and even a welcome end to the existing order, followed by a sense of redemption for those who hold faith in its replacement.

Digital disruption aligned with a broader spirit of the time – a need for renewal centred on new technologies of access. But it also aligned with the grand 'epochal' claims of the period's management and policy rhetoric, which emphasized a need to transform, modernize and do away with bureaucracy, in pursuit of new, more entrepreneurial futures, where work is conditioned by its responsiveness to the needs of consumer markets (Du Gay 2007). Moreover, as Mike Savage (2009) has argued, this kind of 'epochalism' also characterized new approaches to knowledge at the turn of the millennium, particularly a certain 'style of thought' in social analysis: Modernity transitioning to Postmodernity; Fordism to Post-Fordism; a new role for experts in a newly knowledge-intensive Information Economy. Whether in management practice or academic analysis, these kinds of analyses promote a 'narrative conception of temporality which cleanly separates out a past and present' (Savage 2009: 218). Reality is rarely so clean. Turning now to explore three distinct but interwoven expressions of the access narrative, we will need to remain wary of such epochal certainties.

Access to music

The rise of 'access' ostensibly displaces another term: ownership. But what was being 'owned' that is now no longer? The obvious answer is the physical commodity. It is 1982, and you want to listen to *1999*, the new album by Prince and the Revolution. You go to a local record shop or the large music section of a department store, browse the LP section a little, then take the album to the counter, exchanging your cash for two twelve-inch vinyl discs housed in their glossy purple square sleeves. At home, you unwrap and remove the first disc, before placing it on the turntable, turning on the amplifier and moving the needle to the first groove. That album is yours and no one else's – until, twenty years later, you decide to sell it to a local buyer, along with the rest of the vinyl collection taking up much-needed domestic space. Grumbling about how much, or how little, you get for it, the money goes towards a new Pentium 4 processor for your desktop PC, onto which you've been fastidiously downloading and cataloguing your music collection, mostly for free. More time passes. It is now 2022, and you want to listen to *1999*. You take out your phone, open up Spotify and start typing in the search bar, getting as far as '199 . . . ' before Prince is automatically suggested to you. Tap 'play'. Instantly, it starts.

Formats shift. Analogue media, such as vinyl and cassettes, displaced by digital media with a physical carrier, such as CD and the ill-fated MiniDisc, themselves became outdated by digital downloads (MP3, AAC, FLAC, etc.) and, most recently, streaming media services. As many media scholars have pointed out, this was not the uninterrupted movement from ownership to access that is often characterized in popular literature. While the advent of downloading is sometimes reduced to a particular file format, the MP3 standard, this was only one of at least 'four inter-related technological innovations' that David Hesmondhalgh (2009: 59) identifies, alongside the broader take-up of 'flat rate, high bandwidth [internet] connections', 'multi-media computers'; and 'free and relatively easy-to-use software'. To this, we should add smartphone devices and networks that digitized the mobile listening practices of earlier technologies like the Walkman (Du Gay et al. 1997; Qiu et al. 2014). As this suggests, technical innovations depend on deeper infrastructural roll-out and state action – no point in having a faster car with no roads to drive it on – while intangible digital files imply an expanding industry of consumer electronics manufacturing, distribution and retail that is all too physically tangible (Hesmondhalgh and Meier 2018). But even the *de facto* digital standard of the MP3 alone did not emerge overnight. It involved iterations of experimentation and failure, legal confrontation and piratical subversion, while the considerable quantities of time, resources and physical labour invested by various actors – official or nefarious – to effectively intervene in existing processes of distribution were hardly obvious to everyday users (see, *inter alia* Jones 2002; Leyshon 2014; Morris 2012; Spitz and Hunter 2005; Sterne 2012; Witt 2015). In other words, this was a slow, uneven game of catch-up, played out over a number of decades: through the extension and enforcement of legal frameworks, the development and application of technical systems, and transformations in cultural attitudes and consumption practices. It also remade the corporate workplace in ways I will explore over the course of this book.

My emphasis for now is slightly different. Despite the change in packaging and speed of encounter, a consumer's relationship to a musical recording was *always* one of access – at least in legal terms. In 1982's over-the-counter transaction, the only transfer of *ownership* that took place related to the materials that had gone into manufacturing the disc and its packaging – and *not* the music itself. In terms of the latter, what was really being paid for was the restricted right to a private encounter with the record's contents, in perpetuity or until sold on. This right would not, for example, have extended to the capacity to play *1999* at a public event, copy it for a friend, or duplicate it on a cassette tape to play in the car. Of course, the vast majority had a poor grasp on such restrictions and their transgressions rarely resulted in prosecution in any case. But, outside of the most egregious cases of mass piracy, the BPI's claims in its campaign the previous year (1981), that 'home taping is killing music', simply did not stack up – and, moreover, were impossible to systematically regulate (Bottomley 2015). Nonetheless, their fierce response to home taping alerts us to what actually *is* owned – the underlying intellectual property rights – and, moreover, *who* owns it: in most cases, not just the artist but the recording and publishing companies that fund their work. Prince himself had long contested these ownership dynamics, later tying the actions of 'The Man' to a history of brutally racialized property relations: scrawling 'slave' on his face in the mid-1990s he declared, 'If you don't own your masters, your master owns you' (Vats 2019: 118).

Digital formats visibilized such undercurrents by breaking down what, in economic parlance, is termed 'excludability' and 'rivalry', qualities around which the exploitation of intellectual property is modelled. Physical formats are 'private goods': sellers dictate when, where and on what grounds a record or CD is sold (it is *excludable*), and, once gone, it is no longer available to another buyer (a *rival*). But information, such as a digital file, is different: multiple users can download the same track, at the same time and place, with no detriment to other consumers. This makes it closer to a 'public good', like a language. These properties enabled 'home taping' to take place at a significant scale, presenting a 'major challenge' to a *status quo* predicated on 'artificial scarcity' (Hesmondhalgh 2009: 59), perhaps even paving the way to a digital 'commons' (Gilbert 2012), in ways that have been extensively documented. My interviewee Nick framed this challenge in terms of his own role in commercial partnerships: 'you're competing with a product where you can press the blue button and get it for free or the red button and pay for it – and it's the same product'. In other words, this was not just a copyright enforcement issue but a consumer experience issue: how to reinstate excludability by making scarcity *desirable*. As Tony, Legal Counsel for one major label, put it, the industry sought 'new products that are better than pirate products, therefore ones that people are willing to pay for': 'You've just gotta make it unattractive, you've gotta make it boring and we've gotta make the real deal just feel much better.'

This question of 'feel' raises another sense of ownership, in terms of the 'material culture' of goods: the objects that furnish a home, that are gifted from one to another, and which come to organize our everyday sensual experience (DeNora 2004; Hennion 2010; Straw 2011). The symbols, feelings, norms and practices associated with the tangible stuff of music – the smell of the shop, the liner notes, the much-loved paper tickets – are very real, giving rise to meanings, values, affective attachments, investments, memories and collective imaginaries (Frith 1996; Hesmondhalgh 2013a). Unlike the legal rights of the consumer, this material culture *has* changed substantially (Nowak 2016; Roy 2015), with the resistant materials of physical objects displaced by the smooth aesthetic flow of ambient textural moods in the digital 'stream' (Rekret 2019). Notwithstanding the so-called 'vinyl revival' and the YouTube appeal of 'unboxing' electronic devices such as headphones and smart devices (Bates 2020; Hracs and Jansson 2020; Sarpong et al. 2016), this is not simply about packaging. Copyright frameworks do not only regulate the consumption and use of recorded music as property; they actively constitute the listener's practice of accessing the intangible qualities of sonic experience (Kretschmer and Pratt 2009). As such, a crucial transformation lay in the ability to physically encode the abstractions of intellectual property (IP) law into the complex technical architectures that shape musical encounters. For the MP3 download, an attempt was made to engineer such an architecture through Digital Rights Management (DRM) encryption, restricting the use of files and especially the listener's ability to play them on different devices. DRM schemes 'neatly illustrate how intellectual property is really an attempt to enforce a trade monopoly', in Sterne's (2014: 192) words, by reintroducing the capacity to exclude users, more strictly aligned with the copyright rules: 'asserting control over an economy by force when law and custom are not enough'. Applied to the download of individual files, DRM modelled the logic of consumer retail for online transactions – a simple attempt to digitize the record shop – but also made the user's (absence of) rights *after the point of purchase* far more apparent.

As a regulatory tool, this was a blunt one: too explicitly restrictive and open to circumvention. Nonetheless, DRM demonstrated the potential to design legal frameworks into music's information architectures (Burkart and McCourt 2004), wherein lay the opportunity for streaming platforms – and Spotify in particular – to justify their potential to make something 'better than pirate'. Jettisoning the last vestiges of retail and replacing them with the promise of convenience associated with renting, supported by advertising, the platform established the digital walls and doors through which access is controlled. This rendered streamed music not as private, public or even common property but as what economists call a *club good*: non-rivalrous but excludable. Much like a real club, anyone can get in, but only after paying the entrance fee. Moreover, once 'inside' – while navigating the app's various functionalities – we bump up against the regulatory restrictions of copyright in a way that was simply not the case when purchasing and listening to a vinyl record, tape or CD. This becomes apparent when the conditions of access, and moreover, the very content that is being accessed, are changed at a moment's notice: the entry charge is raised; a song is removed or replaced due to a contentious sample or contractual dispute; permissions are rescinded upon moving from one nation's legal jurisdiction to another. And so the architecture of our engagement changes: the club's walls shift, doors are locked or unlocked, the contents of the space are made more or less prominent.

Ultimately, then, the issue of 'ownership' was hardly displaced by the new emphasis on access but rather rhetorically backgrounded: rendered as old news or a behind-the-scenes technicality. For many fans of a previous generation, the loss of music's tangibility – the conviviality of the record store, the gifting of music to a loved one, the 12" vinyl artwork, the placement of needle on groove, the ink left on fingers after reading the music press – is profound. But from a legal perspective, consumers had never really owned anything of economic significance anyway. A change of packaging, from a cassette box or vinyl sleeve to a digital file, streaming audio player and mobile device, made little difference.

Crucially, though, as opportunities around the world grew to synchronize music with other media (film, TV, advertising, video games, online content), and licence catalogues to platforms and 'MusicTech' business intermediaries, what became especially significant was not the ownership of a single song or album but of large catalogues of rights and the metadata attached to them (Watson and Leyshon 2022). That is, issues of ownership became far *more* important to recording and publishing companies over the first two decades of the twenty-first century, not less, and were revealed as primarily an issue of production, not consumption.

Access to employment

A second way in which notions of access reshaped music industries over this period, then, concerns the formalization of labour markets and pathways into creative work. One expression of this is captured by the renovation of cultural policymaking inaugurated by the New Labour government at the end of the 1990s. Key here was an effort to break down long-standing divisions between 'elite' and 'popular' arts, with the new Department for Culture, Media and Sport (DCMS) given responsibility for commercial industries like music, book publishing, architecture and fashion, as well as sports and leisure, alongside traditional cultural domains such as museums, galleries and heritage sites and public service broadcast media. Together, these became the freshly-minted notion of 'creative industries'. The man in charge of the new ministry, Chris Smith, assembled a 'Creative Industries Taskforce' – a group of interested ministers, civil servants and leaders from institutions and business – to define these industries. Their influential 'Mapping Document' (DCMS 1998; see Gross 2020) concluded they were all producers of highly valuable but *intangible* goods and services, with value generated by individual creative talent and captured in intellectual property.

Another publication that same year saw Smith set out his manifesto for *Creative Britain*. Alongside 'creative industries', issues of 'access' and 'excellence' were established as founding principles – that is, commitments to widen public engagement with creativity in the same breath as protecting artistic quality and recognizing great talent – two ideas often viewed as mutually exclusive (Smith 1998). While explicitly framed in terms of audiences, this agenda was very much concerned with remaking economic opportunity and social mobility on a wider scale. Hence 'education' was the fourth pillar, reframed as 'training': good for the public purse; an investment in one's 'human capital' and future reward for individuals. A more open higher education system was one goal, a more creative and meritocratic labour regime was another, preparing the future workforce for fulfilling knowledge-intensive careers. Education and 'employability' in culture and creativity were increasingly the key policy mechanisms for fuelling innovation, growth and international competitiveness across the wider economy (CBI 2010: 4; DCMS 2008: 25; Skillset 2012).

Several schemes were trialled. The 'New Deal for Musicians' attempted to formalize pathways into music careers, giving aspiring artists access to specialist consultants under the broader welfare-to-work training programme – ultimately unsuccessfully (Cloonan 2007: 103–117). For those behind the scenes, Labour set up employer-led sector skills councils, bodies for supporting apprenticeships and training opportunities, guided by employer needs. Of these, the most relevant were Skillset, established in 2004 for the broadcast and interactive media industries, and Creative & Cultural Skills (CC Skills) for the cultural sector (including 'music, performing and visual arts') in 2005. They worked to support schemes that would secure the so-called 'talent pipeline' of young people into jobs, promising to formalize existing training from industry bodies like the Performing Right Society, Music Publishers' Association and Music Managers Forum. The British Music Rights coalition of music publishing and collection society representatives, which had been formed around copyright issues in 1996, widened in 2008 to include a cross-industry membership as UK Music. The latter's basic remit was to strengthen lobbying capacities by creating a single, united voice and facilitating government relationships – within which it, too, took up the access-to-work agenda.

This was not, then, a traditionally *dirigiste* industrial strategy of 'picking winners' for state investment; it was simply a more market-friendly cultural policy. At the outset, music was held up as an exemplar of the new Creative Britain – a risk-taking, talent-investing economic dynamo at home, an export industry and an emblem of British soft power abroad – whereby issues of music education, business development and copyright protection were areas ripe for government support (Smith 1998: 82–83). Nonetheless, the very idea of government-sponsored job training is an awkward fit with a persistent industry mythology centred on record men: those social outsiders, hucksters and hustlers who, in consort with other rejects and unknowns, use creativity and smarts to manage their own economic destiny, together bringing down dominant powers, building an empire on their own terms – and having a good time along the way. Such cultural entrepreneurs were born of popular music's fierce anti-establishment independence and a long-standing wariness of government intervention: the journalist John Harris, who we met in the last chapter, claimed that 'one of the few places where a free market works is pop music'; while artist manager Simon Napier-Bell worried that subsidized pop music would neuter the countercultural fire of youth (both cited in Cloonan 2007: 47). If the 9-5 office job is abhorred by popular music culture, the idea that the state might help you get there is doubly so.

Here lies a tension that resonates through the popular histories, executive biographies and anecdotes in which mythology circulates. Canonically, if 'record men' are anti-establishment, they are also said to fit poorly with corporate life. In major labels, the 'seat-of-the-pants style of their early days only devalues stock price': shorn of the 'freedom to fail' they are eventually 'turned into bureaucrats' (Cohen 2004). While 'plenty of important record men worked for majors' – and perhaps, in doing so, transformed the system from within – Murphy (2014: xi) concludes that eventually 'all record men instinctively choose one: the music or the money'. As such, one way to enter the corporate world is by acquisition. Many of the companies started by such individuals – Russell Simmons' Def Jam, Ahmed Ertegun's Atlantic Records and Chris Blackwell's Island Records – are now established components of the major label firmament (even if their founders later went on to distance themselves from their former enterprises). Another entry point, less entrepreneurial but still consistent with industry mythos, is the post-room. Located at the bottom of the building, the post-room offers a space to get a foot in the door, pay your dues, make contacts and work your way up. From Hugh Meindl at Decca in the late 1930s, to a 'ruthlessly ambitious' David Geffen at the William Morris Agency of the mid-sixties (Barfe 2006: 275–6), to Simon Cowell's stint in the post-room of a 1980s EMI. The latter 'quickly identified a major benefit of working in the post-room':

> he had access to all departments. Cowell was pushy and thought nothing of approaching the great and the good at EMI and asking for a promotion. He had only ever wanted a foot in the door; he was confident that once he was there he could quickly push his way into a better role. (Newkey-Burden 2009: 33–34)

The post-room idea suggests a world behind the glamour, a place of sacrifice and opportunity. In a twist, former Island Records president Darcus Beese describes a similar beginning, as a teenage tea-boy for the label: 'you made tea, you collected people's dry cleaning [. . .] I used to have to go and sit in their cars, wait for the car to be de-clamped and drive back to the office' (Lindvall 2012). This gave him access to founder Chris Blackwell, whom he approached and 'tried his luck'; he too 'soon moved up the ranks'.

The post-room myth is recognized as such and taken with a pinch of salt. As Tony sardonically remarked, 'I think the EMI post-room has probably produced more executives than any post-room in the world!' As a 'foot in the door', it exists on the threshold of the organization as a holding space for those, much like the letters themselves, destined to go deeper inside. Getting a second foot through relies on ambition, self-confidence and a certain amount of tactical nous. Meindl is said to have learnt the ropes of the music industry by opening and reading the company mail; reportedly, Geffen was able to intercept a letter from the university he claimed to have attended, stating that they had no record of him, before it reached his employers. This is appealing insofar as it offers a meritocratic ideal of graft and talent unmediated by the artifice of academic certification. Predictably, we tend not to hear stories from those who never graduated past this stage, remaining stuffing envelopes and making tea. Yet the post-room may now, rather like the record man, be lost to the past. By the end of my research, at least one company no longer employed specific post-room staff to deliver mail (employees, or their assistants and interns, collected it themselves).

These myths tend to obscure some of the more mundane and formalized ways in which individuals have entered into corporate life – recruitment agencies, internships, graduate training schemes – conditioned by the standard tools of human resources and equality and diversity policies: curriculum vitae, application forms and job interviews. The internship became a paradigmatic and controversial form of entry into the industry's workforce, displacing the post-room as an opportunity to show what you're made of. By the early 2010s, unpaid internships were regularly stoking a rich press debate, specifically in corporate music contexts, echoing class-action lawsuits filed against a range of media companies in the United States (Gil 2014; Jones 2012; Lindvall, 2013), as well as a broader sense of internships and free labour at the forefront of a newly precarious young workforce (de Peuter et al. 2015). In response, government raised questions over the hiring practices of what was termed the 'Brit Awards record labels' – companies representing artists nominated for that year's awards ceremony (predominantly the major labels) – 'reminding' them of the consequences of not meeting minimum wage requirements (HMRC 2014).

Strikingly, UK majors were relatively active in developing internship policies that ensured placements were paid. Warner Music's regular raft of paid internships complemented its long-established graduate training scheme; Sony Music developed an 'Intern Academy' to support both hiring practice and on-the-job training; Universal Music had abolished unpaid internships in 2009 and had previously been cited in parliament as a 'leader in the field' (quoted in Lindvall 2013). Nonetheless, such debates indicated a growing recognition that more formal procedures were required to address inequalities for those starting out their careers. UK Music (2014) consulted with the campaigning group Internaware to publish employment guidelines under a general Internship Code of Practice. A broader campaign to improve recruitment practice at a variety of levels across the creative economy led to the development of a Fair Access Principle (Creative Society 2015) and the Creative Access Scheme, supported by the BPI, providing support to young Black, Asian and minority ethnic individuals seeking internships and apprenticeships, as well as the employers who might want to offer them (BPI 2013b). Once again, if quite differently, the question of engineering access was firmly on the agenda.

Access to knowledge

The labelling and defining of 'creative industries' also inaugurated a boom in statistical production. Grouping together the relevant Standard Industrial and Occupational Classification (SIC/SOC) codes allowed this emerging sector to be measured along lines of employment, revenues and contribution to the national purse. A perennial issue, however, has long been that such codes, defined in an earlier era, are notoriously patchy, failing to capture the full range of these industries' activities (Pratt and Bennett 2022). A methodological sub-industry was born, attempting to develop innovative statistical instruments or make novel use of existing datasets – which continues today. In the meantime, government was reliant on industries self-reporting their own data, commonly through trade bodies. In the UK, the key body for music was the British Phonographic Industry (BPI), which represented those firms that paid membership fees and was part of the International Federation of Phonographic Industries (IFPI), providing a route to detailed internal financial data that was otherwise hard to access. Industry reporting from DCMS and BPI was followed by reports from CC Skills (on employment), from PRS ('Adding Up the Music Industry') and from UK Music ('Measuring Music'), not to mention a number of audits of local music ecosystems, provision and closures (Bennett 2020b).

The production of numbers is not a purely neutral exercise; decisions must be made about what numbers to capture and how to display and interpret them. Problematically, both BPI and IFPI represented *recorded music*, not live music or the many ancillary activities (such as retail or instrument and hardware manufacture), and it had long beenargued that they were dominated by the major labels (Harker 1997). Hence, the statistical picture they provided of 'the' music industry (singular) was criticized as, at best, partial and, at worst, an ideological fiction, telling the story of an important industry in rapid decline and in dire need of tighter intellectual property protection and enforcement (Williamson and Cloonan 2007). In addition, the public headline numbers rarely give much indication of underlying detail. As Arditi (2021) describes in relation to IFPI reporting, the full documents typically paint a more complex picture – but only if one pays a prohibitively expensive fee to receive a physical copy. Raw data based on member organizations' submission of data is impossible to access, obscuring methodological changes and glitches over the years and likely supplemented by a healthy amount of extrapolative guesswork. Independent verification of such data is not on the agenda. Although commonly reported in the press and academia as the nearest we have to an objective reality, in practice, these statistics are again characterized by the strict control of multiple layers of access.

Who produces knowledge about whom, and who can make use of it, is then a third theme – and one that bears directly on the research underlying this book. When negotiating access to interviewees and workplaces, I would frequently be referred to HR or corporate communications teams to vet my intentions. One accommodating representative was keen to explore if my findings might either make contributions to a company rebrand exercise or, alternatively, surface employee feedback on HR initiatives – since employees could not be surveyed, I was told, as this would likely prompt discontent followed by an expectation that the company should act on its results. Clearly, my ability to gather information and produce knowledge was conditional. Perhaps for good reason, then, it also generated suspicion among interlocutors: this was a period during which the ethics of insider knowledge were widely discussed, including around the office. The News International 'phone-hacking' scandal dominated headlines in 2011, with the WikiLeaks publication of Sony Pictures emails in 2014 – both held implications for media industry research practice (Connor 2015; Munnik 2016).

Some colleagues clearly viewed me in this light of investigative journalism: eavesdropping and looking for clues that might bring down the organization. Indeed, after WikiLeaks revealed the inflated salaries commanded by their superiors, some clearly viewed my apparent journalistic status as a prompt for them to share, rather than to withhold, information – requiring some judicious care on my own part.

This suspicious, on occasion paranoid, approach to gaining knowledge (cf. Deleuze and Guattari 2004: 123–64; Sedgwick 1997) owes something to the Marxist critical tradition. Marx had located the 'secret of profit-making' in the fact that the capitalist system compels people to seek work while presenting it as a condition of individual freedom: one is free to earn as much as one likes in order to live as one chooses. By contrast, he sought to guide the readers of *Capital* beneath the 'noisy sphere' of consumption and commodity exchange, over the 'threshold' of the workplace, and into the 'hidden abode of production', where the exploitation of labour (a description of the capacity to extract economic value from labour as much as any moralistic sense) takes place (Marx 1976: 279–280). For him, this meant the mill or factory hidden behind the coat in the tailor's shop; for us, it might be the studio or label behind the gigs and records. Later writers working in this 'labour process' tradition have used workplace ethnographies and workers' inquiries to unpick the specific tasks, contracts, conditions of employment and potential for resistance over time (e.g. Beynon 1993; Braverman 1998; Burawoy 1979), including in creative contexts (Figiel et al. 2014; McKinlay and Smith 2009). These have been hugely productive, and I draw on their efforts in this book.

In the fragmented, multiple-contracted, precarious world of contemporary work, however, the abode is not just 'hidden' but distributed and constantly shifting (Böhm and Land 2012). The space of production is rarely bounded within a single building. The secretive and highly relational contexts of media and cultural production make clear that access is not simply a matter of 'getting in'; it provokes researchers to ask: 'Into what?' (Ortner 2010). As anthropologists had long warned, ethnographic access was as much about developing 'rapport' with those being studied as it was simple entry into a field site, posing issues of building relationships and trust, as well as serendipity and opportunism, conditioned by the particularities of researcher identity and fit with the community under study, very often across multiple sites (Harrington 2003; Marcus 1995). To be an 'insider' is typically to work across a number of formal and informal boundaries, continually constructing and remaking them through professional rituals (events, meetings, signings, parties . . .) and discourses (identities, stories, ephemera, gossip . . .). The tradition of 'production studies', born out of United States screen media research, responds to industrial life in terms of these kinds of shifting reflexive work cultures (Caldwell 2008; Holt and Perren 2009; Mayer et al. 2009). We learn much from this work too.

Gaining 'inside' knowledge, then, is about more than a building pass and a stubborn persistence to keep digging for the silver bullet that will take down the powers that be. Within the industry, as we heard in the last chapter, the true 'inside' is in fact located *within* the 'record men' themselves: those who stake a claim as 'the key witnesses and the catalysts'; those who 'know the real stories – who these stars really are, how the breakthroughs really happened'; those who 'saw the potential in its rawest form, negotiated the contracts, signed the royalty checks, and created the hype' (Murphy 2014: xii). Discursive claims of this kind are powerful boundary markers and remain a site of considerable competition over authenticity. As ageing executives busily go about turning their biographies into instructive guidebooks for a popular audience (see Chapter 9), it is the academic researcher's job to do something other than simply read 'the discourse'.

My approach dialogues most explicitly with the Popular Music Studies (PMS) tradition which, in line with the broader trajectory of cultural industries scholarship (see, e.g. Hesmondhalgh 2013b: 37–63), has sought a path between culture and economy, between top-down and bottom-up, and between proximity and distance. PMS' readings of the music industry have typically positioned themselves against totalizing claims of cultural production – with Adorno and Horkheimer the typical pariahs – that saw only the mass production of standardized products, seducing consumers into bland conformity. Scholars went about busily discovering ways in which consumers became 'active audiences', both in everyday life and in innovative 'subcultures' and 'neo-tribes' (DeNora 2004; Fiske 1992; Hebdige 1979; Hesmondhalgh 2005; Jenkins 2006). Meanwhile, in Meaghan Morris' (1989: 21) gloss, issues of work and organization could be sidelined: '"consumption" can be treated as a quasi-autonomous reality diverging from another "reality" called "production" – which, after Marxism, we are supposed to know quite enough about for the time being'. Between these positions, was little room for manoeuvre. Questions of production had often been outsourced to organizational sociology (particularly the US 'production of culture' perspective variously associated with Howard Becker, Paul Hirsch, Richard Peterson and Paul DiMaggio), which had developed more systemic analyses of organizational flows and gatekeeping processes in studies of record companies, TV production and journalists' newsrooms.

PMS is rooted in the tradition of scepticism and suspicion towards industrial organization and commercial exchange. Against established orthodoxies of 'high art' music and the academic musicology establishment, it committed to 'social critique and the challenging of accepted hierarchies', exhibiting an embrace of forms of political commitment that 'may itself have helped to distance [it] from the music industries' (Cloonan 2005: 88). But the century's end marked the emergence of a new analytic approach. Whereas detailed attention to music's industrial production (e.g. Frith 1981) had rarely been backed up by empirical qualitative research on organizations themselves, studies undertaken in the late 1980s and 1990s by PMS scholars such as Negus (1992; 1999) and Hesmondhalgh (1998a; 1998b) changed this. Broadly, they challenged a crude quasi-Adornian view of pop commodities rolling off a factory production line, while also querying whether a more networked, flexible, globalized and informationalized 'post-Fordist' model applied. Their inquiries attended to relationships between transnational corporations and small-scale independents, in particular their mediation through the logics of music genre and political ethos. This chimed with a broader impetus to track reflexive relationships between cultural consumption and production: the design, production, marketing, regulation and identity-making processes around the Sony Walkman (Du Gay et al 1997); the formalization of a new breed of young, predominantly female fashion entrepreneurs (McRobbie 1998); the clubby laddishness of advertising creatives (Nixon 2003); and transformations in technology, aesthetics, organizational culture and leadership, from the avant-garde recording studio to the BBC (Born 1995; 2004).

This 'cultural economy' moment was hugely generative – even if it subsequently reached limits (Negus 2021) – and the theoretical insights and tools it produced shaped the present study. For one, questions of 'access' are encountered here as a methodological challenge to traditional approaches, too closely shaped by earlier modes of production. There are clear practical implications. Researchers must be immersed in the problems that matter to their participants, following controversies and being opportunistic about identifying who, what and where is constitutive of the industry's 'production culture' (Born and Sczepanik 2013). As such, many academics, Keith Negus, Georgina Born and John Caldwell among them, drew on an earlier career in the industry they studied, instrumentalizing the professional sensibilities, as much as the interpersonal networks, that they had already developed. This informed the relationships they developed with participants and the analyses they produced as a result. Reflecting on whether interviewees were telling him the truth or deceiving and misinforming him about corporate music practice, for example, Negus comments that such a question mistakenly 'presupposes that there is some underlying truth about the world, and that we can gain access to this by asking people the right sort of questions in such a way so as to reveal this truth'; and moreover, 'that the person we might be interviewing will be aware of and have knowledge of this truth'. For him, the interview is 'not about "extracting" information or truths that are waiting to be revealed' but 'an active social encounter, through which knowledge of the world is produced via a process of exchange'. This exchange goes two ways. Drawing from the reflexive sociology of Anthony Giddens, he considered himself engaged in the mutual interpretation of the 'double hermeneutic', which emphasizes 'the connections between the vernacular, everyday way in which people interpret and understand their lives and the more technical and formal interpretations and theories developed by social scientists or philosophers' (Negus 1999: 10–12). Industry practitioners, such as artist manager Ben, make use of explanatory concepts and theories from academic settings just as much as researchers make use of the practical everyday explanations of their informants – even if they put them to use in different ways.

This commitment to symmetry is complemented by a more philosophical strand moving into new conceptual territories, among them calls for a 'posthuman' or 'object-oriented' ontology, a 'new materialism' or a more 'speculative realism'. Broadly speaking, these decried the reliance on producing increasingly reflexive discursive constructions as mere 'philosophies of access', reducing underlying ontological realities to linguistic or numerical representations that we can readily understand and interpret. Positivist notions of science as uncovering facts, alongside Marxisant ideas of revealing capitalist forces, were equally outdated; in particular, the 'paranoid' gesture (Sedgwick 1997) of unveiling hidden realities through *critique* had eventually 'run out of steam' (Latour 2004). Dismissing concerns to access 'hidden truths' as anthropocentric, some emphasized the agency of objects and phenomena on a layer of reality beyond what humans can cognitively delve inside, let alone criticize and theorize (most recognizable, perhaps, in the slow unfolding of planetary climate catastrophe); others turned away from theory towards the promise of 'big data' to answer questions we did not realize we wanted to ask (e.g. Anderson 2008a); while scholars in the humanities and social sciences experienced a 'material turn', opening themselves up to a pre-cognitive world of affect and distributed nonhuman agency.

For some, this amounted to a naïve adoption of the practices of post-Fordist techno-capitalism itself: 'a coincidence between today's [realist philosophical] ontologies and the software of big business' (Galloway 2013: 347). Studies of algorithmically-driven streaming, for example (Eriksson et al. 2019; Bonini and Gandini 2019; Seaver 2022), foregrounded the design intentions and effects of globally-distributed computing networks and behavioural data tracking: devices registering not just what we *say* (about our music tastes, for example) but what we *do* and how we *feel*, often without realizing it (what, ostensibly, we *really* like) – and at scale. As such, they called for sensitivity to how humans work alongside complex technical systems with their own agency, attending to the non-linguistic dimensions of unthought behaviours, reactions, passions and affects.

Such approaches contain useful reminders of the need for epistemic humility and symmetry – the sense that theories do not explain everything in advance. In what follows, I am mindful of how productive it can be for our inquiries to attend to and give voice to a whole range of actors, technical objects and sensations beyond the 'usual suspects'.

Yet, rather like the smooth, uninterrupted stream of background music that improves concentration and productivity (*chilled beats to study to*), when 'the desire for explanation' became displaced by 'deliberation upon affect and mood', as the critic Paul Rekret (2019: 66) puts it, '[e]ven philosophy, it seems, was getting into chill mode'. My engagement with industry will, then, focus on the various turns of 'access' outlined earlier, constantly troubling at the boundary of insider/outsider forms of knowledge. But in so doing, as Boltanski and Chiapello suggest, there is a danger that the spirit of critique simply melts into the management of atmosphere (cf. Gregg 2018). Alternatively, as John Williamson and colleagues comment, in a reflection born from the frustrations of engaging music policy communities, perhaps academics just have 'more knowledge of our specialist field than its practitioners – we understand its broader context, we can draw on comparative international and institutional material, we have a longer historical perspective, we have the advantages of disinterest, we are not constrained by encrusted conceptual frameworks'; if so, we should argue more forcefully about 'who really does know best' (Williamson et al. 2011: 471). I return to this question at the book's end.

Outside in/inside out

From around the turn of the millennium, calls to improve access were a central means of understanding disruptive transformations of various kinds. It furnished actors with a sense of why change was necessary and what was replacing it: new modes of consumption, a new creative workforce, new regimes of knowledge production. A range of strategies and scales therefore come to bear on any study of the major record label system at this time, through which it is necessary to resist the simple epochalist sense of a new era – the 'age of access' – naturally emerging in simple transition. A digitally enabled business model of ongoing rents, instead of one-off purchases, did not dismantle the issue of music's 'ownership' but rather increased (and occluded) its significance within a broader political economy. The shifting and opening of pathways into work has hardly done away with crafty entrepreneurial ambition and the importance of informal networks, or indeed with uneven demographic patterning, but instead transposed them onto a new set of institutional relationships of training, recruitment and support. Likewise, if opportunities and techniques for gaining access to the music industry as sites of research have multiplied, the practical and theoretical challenges of working out what exactly is worth paying attention to and how to generate knowledge about it remain as pertinent as ever.

But I should come clean. This threefold concern with music, work and knowledge maps onto my own biography: a teenage concern with how to get hold of music as easily and cheaply as possible, and ideally for free, fuelled subsequent attempts to begin a career in London's music industry, which would itself inform a later academic research study. Yet these experiences also express the complexities and nuances of this linear narrative. My rampant piracy of digital music files and software (*mea culpa*) coincided not only with as many purchases of physical music and production hardware as I could afford but also helped build a sizeable library of music from across eras and geographies - as educative as it was consumptive. It also became a resource for my own creativity, as one of so many amateur bedroom producers, with an ambition to develop my practice and learning through an undergraduate music degree programme. After graduation, the role specifications I encountered explicitly screened against the 'foot in the door' narrative, deliberately puncturing applicants' preconceptions of glamour and instead emphasizing task and process – experience with fast-moving consumer goods; familiarity with bespoke IT systems – alongside the ever-present 'passion for music'. Once inside the workplace, I was naïvely confused to find my commitment to freely accessible music routinely met with shock and derision among colleagues – mostly music aficionados whose passions manifested as formidable expertise, their understanding of popular music history verging on the encyclopaedic. If I had preconceptions that corporate employees might be commercial cynics, there was zero doubt they knew their stuff – and they cared deeply about it.

Consumer and professional identities merge into and bounce off one another; the practice of music work informs and incorporates expert knowledge about its intricacies. In life, as in research, divisions are blurred. And so, throughout this book, I try to practise a form of epistemic symmetry that moves back and forth over multiple thresholds and points of entry. Rather than imposing a predetermined analytic framework, I want to use the three categories of consumption, production and knowledge to travel with the insights, experiences and concerns of those doing the work – not to take them at face value as the 'real' story but as a means of opening up the music corporation from the inside out.

Big Music

The Kensington Record Label Cluster

When I joined, a few people were saying, you know, 'why are you going into the music industry? It's dead'. I saw it as a massive opportunity. And I was passionate about music and I was really interested when I started to find out who these people were, you know, after so much scrutiny and getting such negative [press] . . . And what I realised quite quickly is that it's full of really intelligent people. You know, really smart! . . . I think part of the problem for the music industry is that it's bloody well in the sticks, in somewhere that no-one creative lives! Siege mentality. I don't want to commute to High Street Ken every day. It's on this massive intersection out of London and I always found the pubs, and the music industry's, sort of, after-work culture – it just wasn't there. Of course, we went to gigs. But we didn't really get drunk there. I saw less drug use in the music industry than I've seen in any other . . . But all the old-school, all the old boys, they all still live in West London.

Debbie, Interview (2015)

The 'massive intersection' mentioned by Debbie, a former major record company Director of Insight, funnels drivers onto the A40 Great West Road, one of the main historic arteries linking the heart of London to the West Country. Tracking alongside it is a four-mile stretch of the slightly smaller A315. Starting from well-at-heel Knightsbridge, playground of royals, oligarchs and socialites, it runs along the edge of Hyde Park, weaving past Kensington Palace and the diplomatic embassies of Holland Park, skirting the edge of Shepherds Bush and down through the noisy interchange at Hammersmith to the rather leafier Chiswick towards the city's outer edges. To follow this route is to cut through a web of invisible lines traced by the British music industry's shifting property portfolio. EMI moved to their Brook Green premises, just off Hammersmith Road, in 1995, on departing the iconic central London location in Manchester Square (as featured on the front cover of the Beatles' 1963 *Please Please Me*) – later downsizing into their Wrights Lane building off

Kensington High Street under the hands of the private equity firm TerraFirma in 2012. Along with the sale of the Parlophone label, their building passed into Warner Music's hands, supplementing the latter's Kensington Church Street headquarters just around the corner in 2013. Sony resided on the other side of the same complex (opposite the Derry Street offices of the *Daily Mail* newspaper) from 2010, when they decoupled from BMG to become the 100 per cent-controlled Sony Music Entertainment – consolidating their label offices in Hammersmith and Fulham in the process. Universal condensed its label portfolio upon moving into its Warwick Road HQ in 2005, bringing Polydor and Island Records with it (both abandoning their Hammersmith buildings), and acquiring Sanctuary Record Group – then down the road in Brook Green – shortly after. It expanded into Kensington Village, just off Hammersmith Road, in 2012. It is perhaps unsurprising that the area was christened in one estate agent's council-commissioned report that same year, the 'Kensington Record Label Cluster' (Clack 2012).

Why are we here? A number of reasons present themselves. The area does benefit from, to use the touristic cliché, a 'rich musical heritage'. Nearby Notting Hill brings with it a history of sound system culture and the annual carnival celebration, tracking the influence of Black British music on the country that resulted from post-war Caribbean arrivals. Meanwhile, Chelsea still profits from its associations in music and fashion with 1970s rock 'n' roll, reggae and punk scenes, captured as part of the popular music studies canon in Dick Hebdige's (1979) *Subculture: The Meaning of Style*. This part of town has also been home to the kinds of media headquarters (e.g. the BBC, Disney, Nokia), recording studios (Olympic, Metropolis), and music venues (Hammersmith Apollo, Shepherds Bush Empire, Bush Hall, Earl's Court, Notting Hill Arts Club) that act as key suppliers to large record companies. Also important are the international connections: keep following that road, beyond suburban Brentford, Hounslow and Staines, to reach Heathrow airport within the half-hour, lending the area the near-frictionless global mobility that drives so much of this end of the industry (Watson 2008). Along the way, do be sure to note EMI's former manufacturing and distribution centre at Hayes – now a high-concept mixed-use development dubbed The Old Vinyl Factory, offering opportunities to work, shop and live in buildings with titles like 'Record Store', 'Pressing Plant' and 'The Venue'. The A315 road is, then, not just a 'massive intersection' for traffic but for culture and economy: art and commerce, locality and globality, intellectual property and physical property.

The area's transformation – from transgressive music scenes to industry hub to place-branding initiatives – is expressive of broader trends in urban centres. The idea of a record label-driven 'creative cluster' is a nod to the buzz of business activity that locates competitive advantage in the rapid crossfire of people and knowledge, born of the pressure cooker of close-knit proximity (Pratt 2004). Yet the very existence and authorship of a council report sponsored by a local estate agent is a consequence not of the industry itself but of a culture-led property development strategy that trades on the mobility and liberal-bohemian dispositions of a so-called 'creative class' (Florida 2002). National and city-level administrations across the Global North have long competed to attract and hold on to the kinds of media, tech and knowledge workers that were deemed particularly valuable in the wake of urban deindustrialization – with unintended, often deleterious, effects on affordability, displacing existing populations, rebadging homes as investment opportunities and hollowing out local cultures along the way (Peck 2005). In Kensington and Chelsea, such gentrification only compounded existing processes of social polarization, financial deregulation and changing media representation that have shaped this 'Royal Borough' since the mid-twentieth century (Martin 2005). It intensifies the splits in the area's actual demography – affluent and exclusive; poor and excluded; patterned by inward and outward migration – which, as Hebdige's (1971: 51) discussion of subcultural creativity and class makes clear, have long been present. Notting Hill was, after all, where the activist and Black Panther parents of future artists and repertoire (A&R) leaders Darcus Beese (Island Records) and Joe Kentish (EMI/Warner) had, in 1970, been arrested after protesting the racialized police targeting of the Mangrove restaurant. More recently, home to the wealthy socialites of the popular 'scripted reality' TV show *Made in Chelsea*, the area contains 'some of the most exclusive and expensive addresses in the world [. . .] while the homeless and those in housing need are disenfranchised from the right to the city' (Atkinson et al. 2017: 189). A few years after being labelled a creative cluster, a quite different council report (Dent Coad 2017: 6), this time issued in the wake of the horrific social housing fire in nearby Grenfell Tower, dubbed the borough 'a place where inequality has become a gross spectacle'.

This chapter is about the politics of visibility and inclusion – what gets seen and discussed, and what does not, by whom – within a distinct professional milieu. The remainder of the book dwells on the inhabitants of London's music industry, as seen from the offices of the major labels. Its organizational geography, positioned between local communities and global flows, is not incidental here.

Early in the morning of Tuesday the 9th August 2011, another fire on the other side of the capital illuminated a different part of the value chain and a different dimension of music's moral economy. Three days into the riots that had erupted across the country after the police shooting of Mark Duggan, an unarmed suspect from Tottenham, North London, the Sony DADC distribution centre in nearby Enfield on the city's outskirts was subject to a devastatingly effective arson attack. As the nation awoke, a chaos of image and opinion stood in for sober reporting. Slowly, details unfolded of the near-total damage inflicted on the building. An initial cathartic glee sprung up from some corners of social media, which saw the fire as a just reward meted out to an avatar of global media domination: the Sony Corporation getting what it deserved. This swiftly turned to horror as it emerged that the site was a primary warehouse for the PIAS Group, responsible for distributing independent labels like Domino, Warp Records and Beggars Banquet. It would be these more financially precarious companies, rather than the cash-rich majors, that would bear the brunt of the blaze. Popular feeling, not least among musicians, soon turned to disgust and outrage towards disillusioned urban youth for attacking and endangering small businesses, artists' insecure livelihoods and local communities, rather than the forces of corporate capital.[1] The fire revealed not only that moral-economic judgements continued to appeal to a sharp indie/mainstream opposition but also an unfamiliarity with how corporate and independent production systems are entwined in localized networks of distribution and circulation (Leyshon 2014).

This, then, sketches some of the dynamics animating the urban context in which much of the work documented in this book is remarkably concentrated. Inside cloistered corporate offices, I and my colleagues, interviewees and informants were relatively sheltered from its worst extremes. Prime real estate was sold off, and redundancies arrived in waves as a global financial crisis compounded existing industry woes. But complaints generally concerned the rising costs of rent and a pint of beer, or the diminishing availability of tickets and other perks of the job. To speak, as Debbie and many others do, of 'the music industry' as if it were a homogeneous unit of corporate recording businesses located in one particular corner of West London is, of course, wildly inaccurate (cf. Sterne 2012; Williamson and Cloonan 2007). Meanwhile, her description of the area as 'somewhere that no-one creative lives', implying a rigid boundary around what counts as 'creative', excludes the area's long-standing residents – most of whom are hardly aristocrats. In this book, I am interested in a working community that is, to be sure, profoundly patterned by the contemporary distribution of London's inequalities that impact so much of the UK's cultural and creative employment (Oakley et al. 2017). Their experiences do nonetheless capture how something called 'the music industry' is imagined, hived off from other economic activities, lived, practised and reproduced: its internal moral economy. Over the course of this chapter, therefore, I want to argue with a little more precision that it does indeed make sense to speak of this industry as a singular, cohesively structured and cultured entity.

The chapter's first job is thus to outline a peculiarly rationalized system of production that extends well beyond the major labels, and beyond the sale of recordings and exploitation of rights, to govern the conditions of possibility for anyone wishing to make their career in the wider world of popular music commerce. It is those 'lead firms' mentioned previously – representing the world of what we might call 'Big Music' – that work hardest to regulate this system and set the terms of market engagement. For analytic and pragmatic purposes, I collapse these companies into one fictional entity, 'Sonaverse Records', to be explored over the remainder of the book. Second, I argue that the offices of Sonaverse give rise to a distinct professional culture, embedded in a subcultural self-image that embodies a specific set of norms and values – a music industry *identity* – that again extends beyond the majors themselves. The chapter then moves to define these identity characteristics, in the form of a rising cadre of 'Generation X' executives whose professional ethos was formed over the mid-1990s and, during the period under study, began to displace that of the earlier generation (cf. Negus 2002): those Debbie referred to as 'all the old boys'. Finally, then, I draw from two of my interviewees' self-positioning to consider how ideas of professional longevity and displacement take on specifically gendered dimensions.

Introducing Sonaverse Records

At the time of writing (2024), the major record labels are Sony Music Entertainment (SME), Warner Music Group (WMG) and Universal Music Group (UMG). Collectively, these organizations are commonly known as the Big Three. This has been the result of a slow distillation over the course of the period covered by this book: reduced from a Big Four with the sell-off of EMI Group to Universal and Warner in 2012; from a Big Five when Sony and BMG entered into a joint venture in 2004; and from a Big Six when Universal was formed from the merger of Polygram and MCA in 1999. Together with other large firms in the live and publishing industry subsectors – such as Live Nation, AEG Worldwide, BMG Rights Management and Collective Management Organizations (CMOs) such as PRS for Music in the UK – the majors have an outsized influence in regulating the diverse set of institutions and relationships that make up the system of production commonly referred to as 'the music industry'. For these reasons, we might refer to these dominant members of the ecosystem as 'Big Music'. The analogy here is to the vernacular media framing of other industrial sectors characterized by the oligopolies of a few large but interdependent firms with concentrated market power: 'Big Oil', 'Big Pharma' and especially, given the ongoing integration of rentiership models between music and digital platforms, 'Big Tech' (Birch and Cochrane 2022). Through even larger owners such as Sony Group (owners of SME), Access Industries (WMG's parent), the Chinese Tencent and the French groups Vivendi and Bolloré (major shareholders in UMG), each forms part of an empire of technology, media, telecommunications and consumer electronics, alongside logistics, natural resources processing and real estate investment.

My focus in the remainder of this book is grounded in the UK offices of the majors themselves – namely, during the decade of 2005–15, Universal, Sony-BMG/SME, Warners and the fateful break-up and absorption of EMI. For much of the time, however, I will rhetorically consolidate these further into a single fictional entity: Sonaverse Records. I do so for methodological, stylistic and analytic purposes. First, it helps preserve the anonymity of research participants. Second, moreover, I want to avoid replicating the documentary reportage (and occasional outright gossip) of trade journalism, with its focus on recording the cyclical churn of ownership, sell-offs, investments, market share, corporate restructuring and appointments of key personnel. I will refrain, then, from detailed overviews or histories of the various Big Music companies and their interactions, which are largely recoverable elsewhere (e.g. Eriksson et al. 2019; Hu 2020; Negus 1999; Wikström 2013). Abstracting from the lurid period detail serves a third, analytic purpose, enabling me to focus on the question of how both macro-level 'industry' trends, as described in the previous chapter, and micro-routines of everyday working life are mediated through the organizational form. What results is an emphasis on these companies' similar organizational structures and strategic incentives, as well as production cultures and locational choices.

In making this particular gesture, I acknowledge a significant risk. Narrowing attention on the activities of three already-celebrated companies may, as critics frequently comment, downplay the plurality of music business activities as well as a wider sense of infrastructural dynamics and political economy. Here I am in danger of naïvely reproducing the claims made by corporate communications departments and other representative bodies, such as the BPI or RIAA, to speak and act on behalf of independent artists, live musicians and other groups happily existing outside the 'mainstream' – people who may indeed be actively opposed to the existence of such firms and bodies – as if this were an unproblematically united 'community' with shared goals. In an influential article, John Williamson and Martin Cloonan (2007: 305) argued that repeated references to 'a' music industry rhetorically obscure 'the reality [. . .] of disparate industries with some common interests': sales of physical goods and different bundles of rights, the organization of live performances, marketing and promotional activities; alongside their manifestations in different locations, experienced differently by different actors, with all the conflict and complexity that results (Williamson and Cloonan 2007: 305). Relatedly, for Justin O'Connor (2022), the very concept of music as an industry misdirects policy attention towards issues of intellectual property (IP) protection, entrepreneurship and export strategies and away from a more holistic conception of culture's role in maintaining and giving form to economic life at a more 'foundational' level. Others make a more radical case: Jonathan Sterne (2012) suggests that a relentless focus on the 'monetization-of-recordings construct' hides the fact that contrary to corporate PR, in fact 'there is no music industry'; rather, music-commerce relations are enacted in many different ways and across spaces and scales, from instrument manufacture through to architectural design and extraction of raw materials, all of which deserve equal attention. From a similar position, Georgina Born has suggested relocating attention away from a monolithic view of industry and towards the ever-expanding economic mediations of 'specific properties and potentialities linked to music's socio-material qualities' under what she terms 'musical capitalism' (Born 2013: 51).

For such critics, then, 'the music industry' suggests a coherent and well-defined professional community, set of structures or practices. But this is only a convenient fiction that hides a plural, messy, constantly shifting and much more interesting reality. While this dominant fiction resides, returning to Debbie's formulation, at a 'massive intersection' of peripheral activities and possible alternatives, it is itself an empty container: a space where 'no-one creative lives'. These are useful reminders – although it is worth putting them in historical context. Williamson and Cloonan later noted, for example, that their initial (polemical) framing was intended to signal a changing world post-1999, in which major labels seemed to be under pressure, whereas the addition of live music representative organizations to the trade group UK Music in 2011 allowed them to legitimately state that *'for the first time,* the UK's entire commercial music industry will be represented by one umbrella body' (UK Music quoted in Williamson and Cloonan 2013: 21). The UK case is perhaps the exception that proves the rule since this kind of collective representation is absent elsewhere in the world, notably the United States. The UK Music settlement was itself temporary. Individuals and organizations that depend on live music performances faced an existential crisis in 2020 due to the Covid-19 pandemic, while recordings and publishing saw a rise in digital subscriptions over the same period. Hence, those same representative organizations abandoned UK Music in October of that year to form a breakaway body, Live Music Industry Venues & Entertainment (LIVE). Yet the case remains instructive, since I hardly wish to downplay the real tensions and divergent material interests involved in such groupings. And the broader point continues to be salient: neither the majors themselves nor 'the music industry' label were so easily jettisoned.

I want therefore to hold on to the idea of a singular 'music industry' precisely because it captures something of how Big Music understands and acts on both the *discursive* and the *material* world it is part of – and here precision is indeed required. *Discursively,* it is not just representative bodies that talk in singular terms (misleadingly or otherwise) but also the vast majority of those seeking to make their career in commercial music. This does not imply that they are more 'correct'; simply that they go about their working lives, making judgements and decisions based on an implicit and underarticulated sense of the boundaries of legitimacy: between who and what counts as a part of the world they operate in. This kind of sense-making is discussed throughout the book, but especially in Chapters 4 and 5. *Materially,* 'business as usual' understands the distinct goods and services produced by different subsectors as mutually complementary and, as such, over the course of the 2000s and 2010s, they became increasingly regulated through inter-related data processing systems and rights management infrastructures that shape the conduct of individual performing artists, independent organizations, and others well outside the 'major label system'. It is, of course, quite possible to 'do music' without engaging such industrial structures: amateur bedroom producers who make beats for themselves and their friends; participants in community choirs; free improv musicians who never play the same piece twice with the same group of people; and all kinds of other 'hidden musicians' (Finnegan 1989). Yet anyone seeking to build their career by developing an intense artist-fan relationship (of the kind corporate actors work hard to monetize) will find it hard to do so without, for example, registering their music to gain an ISRC code or collecting performance income. Such individuals will then be engaging with the kinds of digital infrastructures discussed in Chapters 6 and 7 – which continue to grow and evolve. Audio is increasingly captured, recognized and tracked wherever it circulates, whether uploaded as a fan video or played in a (licenced) venue, and stored as data that is then used for processing payments and regulating intellectual property. Those who do not participate in such structures, and do not aspire to, would therefore lie outside this 'industry'.

In brief, then, we will focus on organizations that exemplify (but do not exhaust) the logic of Big Music, across five key characteristics. I am interested in firms that, first, centre their activities around the creation, acquisition and exploitation of intellectual property (IP) rights primarily associated with sound recordings. In this, they are aided by business functions for developing and supporting such activities, alongside, crucially, the capture and administration of any flows of income and data subsequently generated. In order to do so, second, they establish trading relations with a range of other actors and organizations outside the firm and along the value chain: not only writers, performers, recording artists, sound engineers and their representatives, but also the wider promotional activities of marketing, styling, branding, publicity and sales, as well as broadcasters, distributors, aggregators, streaming platforms, pressing plants, collective management organizations and a host of digital intermediaries. Equally, third, they interact with, and in part overdetermine, value chains for different but complementary music goods and services. In the self-conception of UK Music (2013), these cluster around four such 'commercial assets': most prominently the IP associated with the underlying song or work (publishing subsystem) and with managing and styling artists themselves (branding subsystem); and, to a lesser but still important extent, performances (live subsystem). Fourth, they take a leading position in negotiating terms of trade for other cultural and media industries which do not take music as their primary object, from film, television, advertising, video games, visual arts, fashion, podcasting and online content creation right through to the design and manufacture of consumer electronics hardware (audio tools, smart devices), real estate and financial services. And, fifth, through their corporate communications departments and funding of trade bodies, they take an active advisory role for the setting of government policymaking, from domestic primary education and intellectual property law to international trade agreements and the spread of soft power.

This is, then, a 'music industry' produced through a dynamic set of interlocking networks (Leyshon 2014): one that is, yes, comprised of multiple parts, not always aligned, which shift in relation to one another, often coming into conflict. This book barely captures the full extent of these production networks; it has little to say about the day-to-day operations of live music venues, for instance, or indeed organizations outside the Global North. It nonetheless asserts that a distinct and coherent set of industry logics are held in place by the major label as organizational form. In a multidimensional production system, that is, Sonaverse retains a position of overriding influence as a lead firm that sets dominant cultural, economic, media and political agendas. To call this system 'the music industry' is not to endorse the Big Music agenda but to attempt to describe it with care.

Bureau culture

Within this complex global system, many of the key activities take place within the corporation itself, distributed across its local and regional offices. Others are delivered by external suppliers – including recording and performing artists, composers and songwriters, studios, producers, remixers, engineers, graphic designers, stylists, managers, legal representatives, strategic consultants and so on – in temporary or ongoing relationships, encoded in agreements (contracts to supply services, enter into mutual partnerships or licence IP), or in structures of investment and shareholding, either directly or through the larger conglomerates of which they are subsidiaries. Sonaverse's multiple subdivisions function to manage such relationships. As a result, it seeks to exert an influence over as much of the value chain as possible, yoking actors together in concert, playing an outsize role in setting the terms of trade: products to be delivered, financial obligations, territorial jurisdictions, exclusivity periods and so on.

The contracts that emerge are not just arbitrary paperwork. They are also organizational technologies: procedural rules, or 'scripts', that define professional roles, 'prescribing' certain actions and decisions while 'proscribing' others (Akrich 1992). Equally, there are human relationships and tasks that must be managed. The skills required of the label's artist relations team are clearly different from those of the IT department. Such distinctions manifest in modes of ordering the organization along a broad 'enterprise-administration' tension (Law 1994): a conflict between the commercial passions of market innovation and the pursuit of profit that produces contracts on the one hand; the bureaucratic rationality of rules, efficiency and due process involved in enacting them on the other hand. This structures the organizational life of music, just as much as the 'art-commerce' relation inflects the career of the artist (Banks 2007: 6–7). Even so, administration is hardly an inhuman occupation. Even if contractual terms preside, in practice their enactment remains open to interpretation and renegotiation (particularly if an artist is experiencing commercial success and resetting their terms), requiring the organization to respond to and mediate a web of disputes, communications and egos. In other words, the need to translate between enterprise and administration plays out *within individuals* and throughout *the social life* of work, as much as in the formal systems and hierarchies of the corporation.

Contracts order internal operations, then, as much as external relations of supply. This lends corporate life a set of guiding 'institutional logics' with their own path dependencies (Thornton et al. 2015). At least three are important to note here. First, an organizational division exists between central *commercial operations*, encompassing facilities management, HR, procurement, royalty processing and so on, which keep the company running, and satellite *labels* that acquire, develop and promote artistic 'talent'. Second, within labels, marketing logics divide *frontline*, or 'current', historically 'Top 40'-oriented, artists and their repertoire from *catalogue*, which continue to 'exploit' A&R after fading from market prominence. Finally, a functional division separates the client-facing *front office*, comprising A&R staff, marketers and so on, from a more technical-administrative *back office*, updating the systems and processing the paperwork.

This way of ordering the company is not eternal or unchanging. Majors increasingly began offering 'label services' over this period, for instance, fragmenting the administrative, promotional and distribution functions of traditional record deals into smaller piecemeal packages (Homewood 2018). Simultaneously, internal critiques of the '18 month catalogue "rule"' – indicating the promotional window from release after which all repertoire effectively slips into 'catalogue' – were common in a context in which the availability of the digital archive was reshaping consumption habits (e.g. Page 2017), with implications for company routines. Efforts to link product schedules across divisions – strategically coordinating new releases with back catalogue campaigns – did become increasingly routine (Music Ally 2020). But it is hard to institute such changes overnight. Informants indicated to me ways in which the divergent institutional logics of catalogue and label divisions held firm, manifesting in employees' stubborn sense of 'how things are normally done' as much as in product systems that did not communicate with one another. What this suggests is that any large-scale institutional change must contend with multiple competing logics and rationalities, 'baked in' to corporate life in ways that complicate simple binaries of art-commerce or enterprise-administration.

From this perspective of internal organizational structure, I acknowledge that the 'Big Music' framing contains another important risk: of understating important distinctions between the various companies. As Keith Negus (1999: 36) notes, the ownership of each major label – the relative economic significance of the music division within the broader holding conglomerates – has commonly affected the level of autonomy or intervention from 'head office', shaping strategic responses to broader market conditions. There are also some differences in so-called 'corporate culture' and working routines – differences that became more apparent, often difficult, when two companies and their constituent subdivisions and workforces merge. In this context, informants would report to me their own understanding of internal distinctiveness. The sheer size of Universal is often commented on, for instance. After absorbing EMI in a process that was heavily monitored by the European Union's competition and markets authority, UMG became referred to as a 'mega-major' in the trade press, while its seven-story building at the bottom of High Street Kensington was jokingly referred to as 'the Death Star' by those working elsewhere. Its sublabels each had their own subtle character differences, modulated by its roster of artists – Island's affinity with Black British music in the wake of founder Chris Blackwell's connection with Jamaica; Polydor's more pop-rock character; Mercury's catalogue of slightly edgier 'alternative' acts; and the arrival of EMI-Capitol with another distinctive British heritage inflection. Label managers and A&Rs at each of these were incentivized to compete with one another to break new artists and achieve high chart placings, shaping what was commonly referred to as its distinctive 'culture of innovation' – a euphemism, perhaps, for a highly competitive 'up-or-out' work environment. If SME and WMG pursue such a strategy, it appeared rather less pronounced to my research participants, many having worked across several companies, who emphasized more of a coordinated approach – Sony, in particular, being praised (by my interviewees at least) for a relatively flat sense of corporate hierarchy, approachable senior management and supportive HR.

Meanwhile, the former EMI was routinely noted for its heritage as, in legal counsel Tony's words, 'the quintessentially British label':

It was almost like a branch of the civil service in the seventies and eighties. A lot of senior management wore suits and ties, at Abbey Road the sound engineers would all wear white lab coats and ties, and there was this sense of it being a very proper place to be. It did things properly, it was a public company – and it was a bit like going to school: was almost like you had to, at some point, have done some time at EMI in order to really understand the British music industry at least.

Tony, interview (2015)

With the allusion to a 'proper', 'civil service', 'suits and ties' mentality, EMI is allied to a peculiarly British 'gentlemanly capitalism' of patrician public-minded commerce, populated primarily by elite men who probably went to school with one another, invested in quietly maintaining the *status quo* (Miles and Savage 2012). It is, equally, important not to take interviewees' words at face value or to overstate the differences. Tony's rather rose-tinted depiction took place in the wake of its turbulent change of ownership, much discussed in the wider press in similar terms. This reflects a rather nostalgic lament for an earlier time of a more prominent national identity, of the kind discussed by Negus (1999: 65–9) at the end of the 1990s, where traits and stereotypes of Sony's Japanese or BMG's German corporate culture were depicted as being eroded by the tides of globalization – or perhaps 'Americanisation'. A 2013 visit to Universal's Santa Monica head office produced an uncanny moment as I realized that its internal decor, computer and telephone equipment, and even Wi-Fi password were all identical to those in its London branch. It is worth noting, then, that a loss of distinctiveness and creep of cultural standardization – with EMI as an important inflection point – remained an important way in which participants would narrate change to me across all the majors and so indicate their own sense of their values and what mattered to them.

Despite the various calls for their demise, both 'the music industry' and the major record label form stubbornly persist, as a common bureaucratic – indeed *bureau-centric* – structure for maintaining particular commercial flows. It also exists, I want to argue, as a shared 'economic imaginary' (Sum and Jessop 2013), giving meaningful order to working life, decision-making and financial accumulation. In this sense, we can move beyond a single 'corporate culture' towards a broader professional 'production culture' that flows within and between particular organizations (Caldwell 2008; Mayer, Banks and Caldwell 2009). This more anthropological notion of what constitutes 'the music industry' builds on earlier ideas of an 'art world' (Becker 1982) or 'field of cultural production' (Bourdieu 1993) in which the individual artist-genius is embedded, alongside the 'cultural economy' of managerial life (Du Gay 1998; Thrift 2005). Such imaginaries may be instituted in organizational taxonomies – labels, departments, systems – but also expressed in a wider set of discursive tools – from artist interviews and industry conferences to internal meeting documents and government statistics. In this respect, they furnish participants with resources to make sense of how and why they (should) go about their profession.

This is usefully articulated in John Caldwell's (2008) concept of 'industrial reflexivity'. By this, he refers to the intense production by media organizations and individuals of 'texts' *about themselves* (PR statements, electronic press kits, making-of documentaries, instructional manuals, recruitment literature and so on), which are 'read' avidly by those who work (or want to work) in them, as well as some of the most active fans. Those taking part in this reading and commentary, who have the relevant literacy to be able to select and evaluate sources and ascribe value to them, belong to a shared 'value community', as Georgina Born (2010: 199) calls them. The production cultures in which they assemble and circulate are relatively well-delineated. Because such a community is populated by 'practitioners who are intensely aware of the field, or the community of practice, within which they are working, and in relation to which they conceive of their own work' their interpretations are infused with aesthetic and ethical judgements 'about what is valuable, what is less so, what is emerging, cutting-edge, on the way out, or passé in a particular generic space' (Born et al. 2016: 17). As such, we might say that reflexive production cultures form the discursive ground on which a distinctive 'spirit' of the music industry can cohere. In more bureau-centric terms, they articulate a particular 'ethic of office' (Du Gay 2017) that characterizes the ideal-typical music executive – whether trained up through an elite institution, interviewed in the pages of *Music Week*, or gossiped about over an expensive pint in a Kensington pub.

Generation xecutive

There honestly isn't a typical background. I suppose the one thing that absolutely ties every Universal Music employee together and goes through the core of our business regardless of if you're in A&R, marketing or HR is our passion for music. We are also looking for people who are innovative, creative and driven. (Selina Webb, Director of Communications, quoted in Everything You Need to Know, to Get a Job in Music (UMG and Music Week n.d.))

Reflecting on those who occupied the inner worlds of music corporations across his two studies, using a lens borrowed from sociologist Pierre Bourdieu's conception of social and cultural capital, Keith Negus notes the 'coherent class grouping' that his research subjects shared:

> *Recruited into the music industry during the 1960s and early 1970s, most senior executives are middleclass, white males who have received a privately funded education at 'public schools', or attended state grammar schools, and completed studies at university. Their formative experience has been shaped during the era when rock was gaining cultural value, becoming self-consciously intellectual and respectable. (Negus 2002: 512)*

Shaped by bohemian and countercultural values, these decision-makers had similar social and educational backgrounds, cultural dispositions and professional trajectories. Aligned with a broader (if somewhat more diffuse and unmarshalled) 'art school' habitus (Frith and Horne 1987), the values that informed the cultural agendas they put in place (who gets signed and promoted, how and where) were self-consciously 'liberal', critical, even transgressive. Nonetheless, 'they represented, in condensed form, the preferences and judgements of a small, relatively elite educated, middle-class, white male faction' (Negus 2002: 512). Fifteen to twenty years later, this generational cohort, representing what we might call a 'gentlemanly musical capitalism', is recognizable in my own interviewees' references – both provocative and affectionate – to an 'old boys' club' or, less conspiratorially, the 'old guard'. To that layer of senior decision-makers, many of whom were on their way out was added a newer, younger layer of rising executives, belonging to a cohort referred to in the popular vernacular as 'Generation X'. Together they were joining the ranks of what Richard Florida (2002) has named the 'Creative Class', whose values he characterized in terms of an itinerant nomadism, quasi-bohemian commitment to freedom of expression, embrace of identity-based difference and a meritocratic privileging of talent over traditional markers of social status. This was reflected in the social composition of my research participants, while their tastes and dispositions were formed largely between the mid-1980s and the turn of the millennium. Most of the individuals I refer to, though by no means all, were relative elites with university qualifications. Though no longer art-schooled in the same way, arts, humanities and social sciences subjects were still prominent. Their demographic make-up at least partly reflects a widening of that sphere over the latter half of the twentieth century, with increasing numbers, especially of women, but also of non-white and working-class participants (cf. Mandler 2020). There was also a significant constituency from Western Europe and Australia/New Zealand, exhibiting the greater mobility of a post-Cold War world (particularly the UK's EU and Commonwealth membership).

Nonetheless, they too carried with them the tastes and dispositions of their formative moments, now more closely aligned with what Frith (1986) sees as a shift in sensibility towards technologized pop, displacing the rock moment by the end of the 1970s; or, as in later music criticism, a 'poptimist' embrace of artifice over the 'rockist' veneration of authenticity (Sanneh 2004). Gavin, a department head who had been with the same label since the end of the 1980s, gave a flavour of the older end of this spectrum when he recalled working with the likes of 'Holly Johnson, *Blast*, Oingo Boingo, Tiffany . . . Kim Wilde, Sheena Easton, Adam Ant . . . Adamski, "Killer" . . . the swingbeat artists – Bobby Brown, Bell Biv DeVoe, New Edition, Mary J Blige'. At the other end was digital marketing manager Poppy, whose arrival in the early years of the new century manifested courtesy of a post-Britpop digital passion project:

> *I ran a personal fan website about Graham Coxon, Blur's guitar player. And that's how I got my internship – because I had some personal dedication to doing things outside of university. And I think today when we look for interns that's something we also look for – that personal involvement.*
>
> Poppy, interview (2015)

As Poppy suggests, their own entry was often conditional on deploying personal projects, which began as voluntary labours of love before attracting a following, to bring them into contact with established industry players. Once 'inside', they would come to set the internal tone of institutional cultures from middle-managerial positions – making organizational decisions, influencing internal discourses, and performing much of its hiring and firing – in line with their own aesthetic sensibilities and judgements of worth. By this, I do not mean they sought only those whose personal music collections matched their own. Instead they practised a more general form of 'cultural matching', in which hiring decisions are informed by similar values, experiences and modes of self-presentation (Koppman 2012). Indeed, the new cultural intermediaries' expertise drew from the 'omnivorous' forms of pop-cultural appreciation that became 'the norm for the university-educated middle class', with a stock of cultural capital marked less by high-versus-low tensions than studied eclecticism (Warde et al. 2007: 160–1). As such, however, they consciously looked out for displays of the 'right' qualities and dispositions, typically an active, engaged, curious persona – disarmingly coded as a 'passion for music' (Bennett 2018a). This produces a tension, found elsewhere across the cultural industries, between creative class commitments to 'diversity' and 'meritocracy' and the reproduction of exclusionary working environments and production cultures, carefully regulated along socio-normative lines (Gill 2002; Taylor and O'Brien 2017).

The inner child

The late 1990s political PR project of 'Cool Britannia' is typically associated with the marketing of Britpop, an explicitly retrograde tendency in guitar-based music. It overtly countered the chart dominance of technologically manufactured pop imported from the United States with a 'homegrown' nostalgic aesthetic harking back to the era of the Beatles, predominantly produced by (or at least identified with) white men (Huq 2010). The seeds of the more dynamic 'creative industries' agenda, however, are more commonly located in the (post-)modern experimentation and fragmentation of a post-rave coalition of dance music aficionados (Banks and O'Connor 2017; McRobbie 2016), as it transitioned from a loose-knit *subculture* to a more formal and economically viable set of differentiated *club cultures* (Redhead 1997). Facilitated by its attendant drug cultures, rave and dance music has been romanticized for its openness, in which participants, blurring lines of race and class, could abandon or test boundaries around normative expectations of dancing as self-presentation and sexual invitation (Gilbert and Pearson 1999; Pini 2001; Reynolds 1998). Within this context nonetheless developed emergent performances of insider 'subcultural capital', arguments over what was cool or not cool, cut along gendered and classed lines, that determined who was an insider or an outsider (Thornton 1995). This generated a global economy of megaclubs, touring 'superstar' DJs and attendant support personnel, not least in major cities of the UK.

In Angela McRobbie's (2016: 20) description, this broad movement – from 'clubs to companies', as she says – tracked how 'elements of youth culture, in particular those drawn from the energetic and entrepreneurial world of dance and rave culture' were put to work in radically shaping the fracturing, self-promotional and depoliticizing tendencies of the wider creative economy. In this way, access to creative work opportunities remained dependent on being on the 'guestlist', in her metaphor – the door staffed by insider elites. Not just a scene, it was a cross-disciplinary production culture with its own lifeworld connecting music, fashion, design, fine arts, style publications and promotional industries, busily constructing new consumer categories and lifestyles, such as the 'new man' (and the backlash of 'new lad') (Nixon 1996). In the various 'subcultural outposts in journalism' – such as *The Face* and *i-D* magazines – which 'declared their alliance with the new politics of identity, and its history in the proud allegiance to "style" in mod culture: the term "face" was mod slang for a "top boy", one of the élite, the stars of the scene' (Gilbert and Pearson 1999: 25). As much of the language suggests, this was a particularly gendered discourse that set the scene for an emergent 'post-feminist' sensibility, in which female equality was declared to have been achieved, rendering feminist commitments a thing of the past (Gill 2003). This all came together in increasingly expensive inner London areas such as Hoxton (Pratt 2009) or Soho (Mort 1996; Nixon 2003), assembling a loose-knit professional milieu of artists, designers, intellectuals and eccentrics, who gathered in media agencies, bars and – perhaps especially important – the new Bohemian private members' clubs such as Soho House or the Groucho Club (Edrisi 2020). This dual 'club culture' suggests how access and belonging to London's creative class, regulated through style and insider status, moved from the scene to professional life.

And yet we should also take care with this narrative, which is not itself evenly patterned. To illustrate this, I turn to two career trajectories. Andrew's biography is exemplary, beginning with his own time on what he called an 'ultra-critical' undergraduate programme.

I wrote a dissertation on music and subcultures and got a 2:1. This was mid-nineties. Always loved music, always loved culture. But I didn't really understand the real blessing that was gonna come with going to [university], which was being in an ultra-creative, ultra-critical environment, with those kinds of people, and spending formative years of your life living in a socially deprived area of London. It really was a mixture of different types of education, in the kind of strict academic sense, in the subcultural sense, and in the forced development of social skills and emotional intelligence that come with being pulled out of an entirely white, middle-class, West Country environment, and plonked into a much more challenging, culturally diverse part of a big city. I was actively involved in music – promoting nights, DJing, that kind of stuff. [. . .] I was particularly keen to start my own record label and to really invest more time in what I felt passionate about, which was a, kind of, underground subcultural side of dance music. [. . .] I wouldn't say that at the age of seventeen or eighteen I knew I was gonna end up in the music industry but I knew I loved music, I knew I wanted to understand more about the way in which people interacted with one another – that's about it. And I knew I wanted to go somewhere hip(!), to hang out with like-minded people.

Andrew, Interview (2014)

Andrew is reflexive about his social advantage, contrasting a private-school upbringing with the urban milieu in which he took his degree. His repeated references to the 'subcultural' character of his own tastes and activities signals – in part, presumably, for my benefit – a flourishing of authenticity credentials, alongside knowledge of cultural studies texts that would have been encountered through the curriculum (Hebdige 1979; Thornton 1995). But his intellectual stimulation came less from his studies than from fellow students. He views education primarily as an opportunity to 'make sense of your life', emphasizing character formation, divergent thinking and exposure to mixed urban communities. This is indicative of a broader pattern in which, as Kate Oakley (2013: 25) writes, 'the experience of going away to college, full student grants, and the chance for a period of cultural and personal experimentation, were all more significant in terms of producing cultural workers than the provision of particular courses at universities'. Defining himself as a music fan with a sociological inclination, and downplaying any calculative intent, Andrew simply sought to 'hang out with like-minded people' – even if his use of an archaic word, 'hip', delivered with an ironic inflection, indicates an awareness of the connotations of being a superficial 'scenester'. Eventually, after ten years as a jobbing DJ and promoter while picking up work in music magazines, labels and distributors, 'I did start to feel like I was getting a little bit old for it', he says. As such, he took a 'very specific decision' to take on a full-time role, having paid his dues on the DJ circuit, and 'move into a more general management type area', more apt to his age and experience.

Andrew appears to embody, almost to the point of caricature, McRobbie's 'clubs to companies' trajectory. His is a DIY career of, in a common trope, making it up as you go along, flexibly putting various aspects of his biography to work for him (Adkins 2013; A. Bennett 2018; Gill 2009). Situated in business affairs, his Sonaverse colleague Sophie's entry ticket also came via the university in the form of her legal qualification. But, by contrast, she described how the major label was less a form of settling down than an escape from the dull routine of being a solicitor: 'I don't feel it's quite the same as working within a law firm, thankfully' – even if 'I thought it would be a bit more *"creative"* . . . I dunno how!' – 'I feel like I'm part of a label'. She emphasized the need to adapt her legal knowledge to the new situation, the rapid pace of life and the personalities that went along with it, also 'making it up' as she went along: 'When I started I didn't have a clue [but my boss] was like "well I don't really know anything anyway, we're both learning!"' Both Andrew and Sophie, in their different ways, appealed to their capacity to adapt and be flexible to their situation in ways that revealed both a lack of explicit training and an anxiety towards ageing in a context predicated on youth. In an offhand aside, Andrew hoped his skills were transferable for 'when I get sacked from this job'. Sophie, however, was more explicit:

> *[You] get to your, use-by, sell-by, whatever date? It's just inevitable. As long as I'm wanted in the label, I'll be around. But I think, as somebody that's been here for as long as I have, if not longer, said, 'you should always have a plan B! And a plan C, and a plan D . . . ' Because, you know, you see people that I thought were gonna be at the label forever – and then you see them walking out in tears and being told that they don't have a job any more. And I just think that's part-and-parcel of working in an ever-younger, cooler, moving . . . You know. [. . .] I think there will come a time where, yeah . . . [clicks tongue] But you know, I could be wrong.*
>
> Sophie, Interview (2014)

No one is indispensable; redundancies are 'inevitable', escaped only by chance or voluntary exit. This existential certainty, that the end is coming without warning, stamps employees with a 'use-by' date: a predetermined future point at which their 'product' – their youthful authenticity and 'cool' identity, rather than their bureaucratic expertise – will have 'expired'. Nonetheless, this dynamic climate of workforce renewal appears *culturally* legitimate to Sophie, since the company's very viability is tied to the age of its occupants: a successful business must have a climate of perpetual renewal.

Responding to this context, each interviewee positioned themselves slightly differently. Sophie continued:

> *There are people there that could potentially be my children now. I'm not that old! But you know, that's just the natu . . . – Cos the labels get younger, that's the whole point. They want young, fresh people on board. They don't want a load of dinosaurs. [. . .] I think the few of us who are the old guys are still quite fun! [. . .] You know, as much as our hearts and our hangovers . . . ! We still are part of the*

team and I consider myself someone that people can go to and get advice from and just kind of help everyone because I've been here for so long that I just know stuff. [. . .] I've made myself someone that can be approached and can be there to help and to advise and to guide. Especially, as I said, the people who've just started the job and don't really have a clue about anything.

Sophie, Interview (2014)

While still 'not that old', Sophie is now one of 'the old guys', approaching 'dinosaur' status, but defends herself as still 'fun'. Conceding that her energies and resilience (her 'fresh'-ness) may have ebbed, she resolves her anxieties by explaining Sonaverse's obligation to preserve a youthful cultural legitimacy: 'that's the whole point'. In this climate, she is 'there to help and to advise and to guide' – although later acknowledged this was not actually in her job description – portraying herself as a supportive, almost maternal figure, especially for untrained younger generations. There is an indirect 're-traditionalization' taking place, where her conscious decision to trade in the grey conformity of a law firm for the clubby creative vibe of music has nonetheless been shaped along traditional gender lines (Adkins 1999). As such, Sophie might recognize herself in Banks and Milestone's (2011: 80) description of female new media workers: 'required to provide nurturing, emotional support for disgruntled and marginalized team members', acting 'as an unofficial mother figure to "the guys" or, when required, as the soft face of management'.

Reflecting on his decision to switch careers, however, Andrew happily expounded on why precisely he had, in his words, 'survived' in quite different terms:

I'm absolutely delighted that I'm still in the business; it's not an easy business to stay in for a long period of time. [. . .] I love the business, I'm no less passionate about it now than when I was eighteen years old. [. . .] A successful professional in the music industry isn't fussy: that is one of the keys to longevity. The fact that, four or five years ago, the first day I was ever exposed to Spotify by a sixteen-year-old work experience person in my department – I was running around the office like a kid, I was that excited. It's precisely why I've survived as long as I've survived. Precisely why that kid is going to have a successful career in the industry. I'm not fussy about how I get it: I still think like a music consumer.

Andrew, interview (2015)

If Sophie spoke of fading energies and being one of 'the old guys', Andrew makes no such concession. Having retained an 'unfussy' flexibility and thinking 'like a music consumer', he presents himself as a 'passionate', 'excited' teenager at heart, with an itinerant childlike energy that confounds his earlier claim to feel too old for the DJ circuit. But such autobiographical justifications reveal less about Andrew himself than an environment which enables some, not just to 'survive' but thrive, while others encounter seemingly 'natural' limits (Adkins 2013). Sociologist Andy Bennett (2018: 2) has described a growing 'post-youth' dynamic in which 'the leisure and lifestyle practices once firmly associated with people's teens and early twenties [. . .] are now increasingly assuming a more vocational dimension in the everyday lives of those who invest in them'. But this can be turned on its head: it is the major label that provides an apt environment for Andrew to continue feeling 'like a kid'; while Sophie, more closely bound into an administrative role, explains her longevity in terms of acting like a mum. Her capacity to 'reach back' in order to 'move forward' – that is, to draw on the latent resources of her youth – is far less clear. Crucially, however, she is just as willing to defend it, making a 'business case for diversity': that the market is better served by a diverse – but crucially, *young* – 'talent pool'.

All change?

In 2018, everything changed again. Universal Music's recording and publishing arms underwent further consolidation, relocating to North London's newly developed King's Cross 'creative cluster' in 2018, alongside Vivendi's SeeTickets, PRS, Google and numerous other media and fashion companies. The Covid-19 pandemic made further irreversible changes to the geography of the office. Sony followed suit in 2022 with a new site around the corner to Universal that sought to transplant the look and feel of its Derry Street building, while incorporating flexible designs that capitalized on working-from-home practices and an interior layout that echoed a department store, marking the transition between label divisions with distinct brand identities (Mutrie 2022). The move coincided with the close of the period under study in this book. That particular iteration of Big Music centred on a part of the city scorned as a travel blackspot, ill-served by the slow and infrequent Circle Line tube, that was prohibitively expensive to live and work in. Notwithstanding its heritage, and aside from those aforementioned venues, a handful of record shops and occasional star-spotting, present-day

Kensington was seen as a silent vacuum for music culture at the end of a lengthy commute, with few after-work options.

Nonetheless, this concentrated geographical location and position within a broader, interlocking system of production aligned with the major labels' shared organizational structure and reflexive production culture, producing a relatively unified – if internally contested – form that I have christened Sonaverse Records. Its employees' experiences were patterned by internal organizational contradictions as well as cultural concerns inherited from beyond the threshold of the building. The precise location was less pressing for those creative or boardroom elites whose day job involves travel to attend meetings, signings and gigs, for example, than for office assistants with work patterns determined by their ability to reach a desktop computer. If the waves of real estate consolidation served ambitions to bring creatives into new shared spaces that would fire inspiration, it was far more about streamlining apparent 'administrative bloat'. While workforce expansions and contractions can be traced, the question of whose departure led to a new position in the building across the road, and who never returned, is far harder to see. It is clear, nonetheless, that the old guard were moving on, signalling sunset days for the gentlemanly mode of musical capitalism and private members' clubs, in line with a broader shift towards a more 'liquid managerial elite' sustained by business education, trade news and global industry conferences (Davis 2017; Thrift 2005). Through such changes, no doubt the production culture will be remade as a consequence. A new iteration is emerging, *perhaps* embedded in a more 'Millennial' outlook, *perhaps* more attentive to issues of 'inclusion', 'diversity', popular feminist and anti-racist campaigning, with a sense of sociality shaped more profoundly, *perhaps*, by online interaction.

But they are still, of course, in the capital city.[2] The fact that decamping four miles north-east of Kensington seems so disruptive is itself indicative of just how close-knit this world of industry insiders is – even if, as the move suggests, it is more open to continuing sectoral convergences between music, media and tech companies, which remain ensconced in the contradictory transformations of the creative city. For workaday music employees, the utopian creative economy promise of reconciling work with pleasure remains embedded in an uneven geography and entwined in unequal company hierarchies, expansions and contractions, strategies, incentives and pressures, ebbing and flowing with the tides of global musical capitalism. In that sense, there is much that will persist.

Part II

Inside out

4

Re-evaluation

The money trench

The music business is a cruel and shallow money trench, a long plastic hallway where thieves and pimps run free, and good men die like dogs. There's also a negative side . . .

– Hunter S. Thompson

This quote haunted my time in the industry, reappearing in press reports, executive speeches, social media posts, informal conversations and email signatures. Its sentiment is clear and expressive enough to have acted as an epigraph for at least two books dwelling on turn-of-the-century decline: journalist Louis Barfe's acclaimed popular history, *Where Have All the Good Times Gone?*, published in 2004; followed, four years later, by the novel *Kill Your Friends* 2008, a violent romp around the industry of late 1990s Britain by the former London Records A&R-turned-writer, John Niven. Both take cues from a broader sense that an industry in freefall is merely receiving its just desserts after a decade of drowning in what Hunter S. Thompson warns was a zone of hedonism, greed and a visceral contempt for 'good men' who follow the rules. And then comes the twist – where any note of caution is mischievously deflated by the seductive allure of depravity: 'there's also a negative side'. Referencing Thompson's most famous work, Niven suggests that 'fear' (of signing the wrong acts) and 'loathing' (of others' successes) are 'the dominant emotions within A&R culture', labelling the quote 'the music industry's magna carta' (Niven 2008). As a pioneer of the raw and risk-seeking gonzo style, his legacy is felt in the narratives of decadence produced by many 1970s and 1980s rock critics, from Lester Bangs in the United States to Nick Kent in the UK. Writing from a first-person perspective in which the journalist is as much participant as observer, the attraction of gonzo derives not just from its detailed depictions of excess but also from its ethic of immersion and personal commitment – whatever the costs – rather than studious detachment (Mosser 2012). Gonzo communicates a feel for the darker side of desire, intuitively understood and manipulated by the shrewdest record execs, enshrined as part of an authentic 'A&R culture'.[1] It's something of a shame, then, that Thompson never said it.

Or didn't quite – he did produce an almost identical line about 'the TV business' in 1985, which missed off the crucial payoff about the 'negative side'.[2] The adaptation appears to have been popularized through countless online obituaries in the wake of his death. Still, it's a great quote – and if its widespread meme-ification is a poor reflection on journalistic fact-checking in this period, it is nonetheless a testament to Thompson's influence on popular music history, driven by his perceived rockstar status (Mueller 2005). Accurate or not, this widespread circulation within the reflexive production cultures of commercial music tells us something about the enduring appeal of images of debauched lawlessness to industry actors, even through straitened times. For this was the low point of the mid-2000s, with memories of the 1990s 'good times', and the traumatic shock that followed, still fresh.

Keith Negus' (1999) observation that, just as *industry produces culture* (music is a manufactured and marketed cultural product), so too does *culture produce industry* (popular music production processes are textured with inhabitants' tastes, values, habits and identities) is oft-cited. It is a simple but powerful analytic reversal – and equally one formulated during those good times. It was also, I want to suggest, implicitly embraced by major labels themselves during their subsequent fallow period, as part of a strategic shift in identity communication: putting the organizational culture, rather than the failing business model, on public display. I do not suggest that CEOs were taking strategic direction from popular music studies; rather, insiders were anxiously reacting against what they saw as public misconceptions of their moral purpose. Insight director Debbie recalled having her own perceptions challenged after joining Sonaverse in the mid-2000s, as 'the music industry got really slagged off' in popular and press commentary:

> As I was working there, I think there was a shift [. . .] because it was acknowledged that it is a really difficult challenge, how to monetise content on the internet. [. . .] So, I think to a degree people have backed off a bit. And I think that coincides with the industry doing much more [. . .] and participating in debates that it just didn't before.

> Debbie, interview (2015)

Over the subsequent years, as a result of this 'doing much more', the nature of industry culture became a much more visible site of critical contestation and re-evaluation within the popular imagination.

To interrogate this, I first consider how the recording industry appeared in two popular representations (the *Idol/X Factor* franchises; the aforementioned *Kill Your Friends*); second, across two organizational subdivisions (market insights; corporate communications). Through such symbolic and organizational mechanisms, the industry came to see itself as more reflexive, transparent and communicative. Through a series of 'critical encounters' between audiences and employees, to adapt Negus' refrain, the evaluative judgements of industry practitioners became more visible to consumers, at the same time that those of consumers were becoming more visible to the industry. Equally, while debates raged over expenditures for breaking risky new acts (cf. Cooke 2013), so too did the parties and perks, and especially the rhetoric of industry excess, which became far less defensible. Moving away from the unruly 'cruel and shallow money trench' of industry legend, a new professional ethos was emerging, set in relief against a backdrop of diminishing cash flow. Nonetheless, the myths and legends of popular lore – and, in many cases, the behaviours – endured. Consequently, these representations set the scene for new tensions, where the experiences, values and beliefs of those 'inside' the industry culture clashed with the perceptions of those 'outside'. In light of this re-evaluation, Sonaverse's cultural legitimacy also became contested by employees – whose working lives I turn to in more detail in the next chapter.

The public vote

As Matt Stahl (2013: 26) observes, 'in moments of transition or crisis, the rules of the system, both ancient and futuristic, come into dramatic focus'. In the post-millennial popular imagination, perhaps no television show brought these rules into a more dramatic focus than the *Idol* and *X Factor* franchises – while no figure came to personify the British music industry like Simon Cowell. Repackaging the talent show audition, combined with a rags-to-riches reality narrative, the format converted the mythical 'x' factor, an unnameable, unmeasurable and unpolished gift, into the globally-recognized *X Factor* (Wall 2013), a hugely successful brand that produced reliably high viewing figures and chart placements.

The first episode of UK *Pop Idol* was broadcast in October 2001.[3] It introduced Cowell, alongside other judges – producer Pete Waterman, publicist Nicki Chapman, radio DJ Neil Fox – and the concept. A tongue-in-cheek how-to guide to pop stardom – continuously change your image, time record releases with tabloid fodder such as paparazzi photo splashes or acrimonious public arguments, waste money, lose touch with reality and make ridiculous demands – was followed by advice from established 'idols': believe in yourself; prepare for criticism; work hard because there's always someone else. With the appearance of Gareth Gates, the eventual runner-up of the series, whose debilitating stammer disappeared when he started to sing, it also introduced what was to become a central theme of the format over the following decade: the presentation of popstars as real, everyday people battling with a personal struggle.

The format was broadly split into two stages. In the early auditions, the full range of the general public's enthusiasms and delusions is put on display, to infamously humiliating effect, occasionally interrupted by moments of genuine awe. Then come the elimination rounds, in which the cream of the crop performs for an audience each week, before being put to a public vote. Much of the drama is provided by the judges' feedback, often leading to internal disputes among the panel and to public rebuke. The eventual winner would be rewarded with a prestigious contract with co-producer Simon Fuller's 19 Recordings. In 2004, after two series', Cowell split with Fuller to develop *X Factor* – an uncannily similar show with ownership held by his own company, Syco, a Sony subsidiary – which, alongside his continued role on *American Idol*, cemented his global reputation.

As Tim Wall notes, the programme made the A&R process – of 'finding the next generation of singing talent and selecting the songs that this talent will sing' – more visible at the same time that it had become less lucrative:

> When the main commodity, the record, is the product of this process, it does not need to be in public view; but now, records themselves no longer make as much money, so the twenty-first-century solution is to make the A&R process the content of a new form of engagement and the show a new commodity to be sold to television networks, themselves trying to come to terms with the challenges of new media forms. (Wall 2013: 20)

As such, *Idol/X Factor* combined two innovative features. First, it put the execs on display for a mass audience. But rather than simply making celebrities out of particular individuals, it proposed to reveal the professional practice of developing 'talent' for a mass market audience. In this, it is one of a number of shows – from *Antiques Roadshow* to *Dragon's Den/Shark Tank* – that, as economic sociologists Fabian Muniesa and Claes-Fredrik Helgesson (2013) observe, treat valuation as a public spectacle, capitalizing on the voyeuristic pleasures of witnessing usually hidden processes of market judgement in action. The television audience is taken inside the decision-making process, demonstrating the deeply sensual and deliberative practices of appraising quality – 'watching, listening, tasting, smelling, touching [. . .] debating, hesitating, comparing, sorting, ranking and quantifying, imagining and inquiring' – and opening it up to the public (Muniesa and Helgesson 2013: 120). That is, industry insiders and their previously hidden processes of expertise are rendered for popular consumption.

The second feature came from the other direction: leaning into the popular interpretive judgement of so-called 'active audiences' – that is, consumers who are recognized to bring their own interests and experiences, who question dominant narratives, engage in dialogue and criticism, produce new meanings and sometimes use these to create entirely new worlds. The decision-making process was thus opened up to the 'semiotic democracy' (Fiske 1992) of a viewing public. Across multiple media platforms – several broadcast shows and spin-offs, major label promotions, radio play, brand partnerships, much later social media – the show encouraged a continual critical evaluation of performers, their performances, the judges' commentaries and the contestants' responses to criticism, ultimately letting viewers cast the final vote. For the media scholar Henry Jenkins (2006: 62), *American Idol* exemplified an emergent 'affective economics', which sought 'to understand the emotional underpinnings of consumer decision-making as a driving force behind viewing and purchasing decisions'. In so doing, it was belatedly catching up with the previous decades' worth of cultural studies' insights into fan communities, absorbing them into established marketing frameworks.

More immediately, the format produced hits. *Pop Idol*'s first series winner Will Young and runner-up Gareth Gates produced the two bestselling singles of that decade (OCC 2019), while acts such as *American Idol*'s first winner Kelly Clarkson and *X Factor*'s One Direction achieved enormous worldwide success in later years. The regional variations spawned by the two franchises are too numerous to mention. As Wall suggests, therefore, this was not transparency and dialogue for its own sake but a means of (re-)gaining some control over the critical encounter between firms and audiences. Cowell's sense that the public had chosen the 'wrong' winner of *Pop Idol*'s second series indexed some tension with this approach. This, as well as the opportunity to bring its acts under the twin music and television production outfits of his Syco umbrella, rather than Fuller's 19, reportedly motivated his decision to develop the *X Factor* brand as a competitor. That show's introduction of an intermediary residential mentoring and development phase altered the judges' evaluation role, increasing the reality television portion of the formula. But its timing in the lead-up to the crucial Christmas gifting market, in which multiple artists traditionally competed to achieve the peculiarly British accolade of Christmas Number One, was equally important. Both shows generated promotional retail buzz for 19 and for Syco, but also for the industry as a whole – or Sonaverse's end of the sector at least. The show was not just about marketing artists, in other words, it was about reconstructing favourable market conditions.

Unlocking A&R

What kind of music do I like? Incredibly, you really do get asked this from time to time. [. . .] Asking a major label A&R manager this question is like asking an arbitrageur what kind of commodities he likes. Or saying to an investment banker, 'Hey, what's your favourite currency?' I have very wide musical tastes. 'Eclectic', as spastic musicians say when they're trying to sound clever in interviews. I don't care which genre something comes from – rock, trance, hip hop, Bulgarian fucking heavy metal – as long as it's profitable. (Niven 2009: 12)

During the same period, author John Niven set about telling another story. Set in 1997 and published in 2008, *Kill Your Friends* (*KYF*) paints a picture of an empire wallowing in decadence. Much humour is derived from its off-hand references to the enormous sums and considerable hype floating around acts fated to become a minor footnote in pop history – while signs of an impending implosion glimmer hubristically in the background. Within this context, the very title of 'Artiste and Repertoire' (A&R) struck Niven (2008) as 'a genteel hangover from another era, reeking of civility and respect, of Ahmet Ertegun in a spun-silk suit leafing through some sheet music with Ray Charles' – in stark contrast to his own recollections of 'a generation of twentysomethings blasted to the gills on cocaine, tearing around the world trying not to lose their jobs by doing something crazy. Like actually signing a band'. *KYF*'s despicable but blissfully ignorant protagonist, Steven Stelfox, is interested only in getting rich by signing whatever appears to be on the verge of commercial success, in order to spend huge sums on drugs, sex and sating his homicidal tendencies. Other people are collateral. After a year of industry life, his junior assistant, a fresh-faced indie music enthusiast:

> *looks like fucking Methuselah now: his skin dry and flaking; his eyes bloodshot and sunken; his hands forever trembling as he lights a fresh Silk Cut with the butt of the last one [. . .] nursing a constant hangover and a three-gram-a-week habit [. . .]. His glossy mane of tyre-black hair is already streaked with grey fissures. He has just turned twenty one. (Niven 2009: 33)*

This is resonant imagery, tapping into a history of anxiety that rational decision-making is undermined by 'animal spirits'. If it is not too fanciful to take a brief detour from the 1990s A&R office into the agora of classical antiquity, for instance, we might note the striking resemblance to the 'horse of passion' described by Plato: 'crooked, [. . .] short necked, snub nosed, black skinned, gray eyed, bloodshot, a comrade of wantonness and boasting, shaggy about the ears, deaf, barely yielding to the whip and goads' – the unstable twin of that sturdy 'comrade of truthful opinion', the horse of reason. Both are shackled to the chariot of the intellect, the philosopher says, requiring considerable care and forceful intent to stay on track. It is a short journey from being driven by passion, he suggests, to a state of obsession, compulsion, exhilaration, addiction and even self-annihilation (burnout, career failure or worse). Skirting dangerously close to that point, Stelfox's strategy is to cultivate cynical distance. Taste for him is not an expression of personal values but a tactical weapon for netting naïve acts, deluded into thinking music is somehow important. Much time is spent attempting to find methods of exploiting the stupefied masses. Drawing on a 'golden rule of showbiz', he sees in the 'outpouring of collective emotion among the lower classes', such as the death of a princess, primarily a sign that 'there are records to be sold and money to be made' (Niven 2009: 221). Colleagues are competitors, and women are resources to be used only to fuel his own energies and career trajectory.

Philosophical allusions aside, the novel is hardly a literary classic. 'American Psycho meets the X Factor in an orgy of mad, gleeful nastiness', wrote one reviewer, 'blackly lampooning the stupid, hypocritical world of the music industry' (Housham 2009). Brett Easton Ellis' satire of 1980s Wall Street, viewed through the eyes of the psychopathic investment banker Patrick Bateman, provides a clear template. So too does gonzo journalism – ironically at a time when that ethos had dwindled in the mainstream music press, driven by organizational changes in magazine ownership (Forde 2001) – as well as the club culture novels of 'chemical generation' writers like Irvine Welsh (Redhead 2000) and 'new lad' magazines like *Loaded* (Nixon and Crewe 2004). Why read it at all then? Because, while *Kill Your Friends* is written for a general reading public, Niven's many references to people, places and situations recognizable to those familiar with the British music industry ensured that the book was also passed around as a key text by many of its occupants. A comparison with Imogen Edwards-Jones' (2009) *Pop Babylon*, a less gratuitous account of the cynical manufacture of a fictional boy band, is instructive. While prefaced with the assertion that 'all the following is true', based on the testimonies of nameless 'insiders', Edwards-Jones herself is a mere bystander, and *Pop Babylon* is just one in a series of similar books with a service industry theme. No gonzo to be found here. Accordingly, at least as far as I saw, the book gained few admirers within the professional world in which it was set.

By contrast, Niven writes in the tradition of the *roman à clef* ('novel with a key'), a genre of writing that has long preyed on the public's fascination with the salacious details of elite social circles, while being dismissed by high-minded critics for falling somewhere between memoir and gossip column. Casting the reader as a voyeur, for whom allusion and speculation are enticing enough, much of *KYF*'s appeal (and promotional advantage) derives from Niven's own history in the field, construing a dual insider-outsider readership of those who know and those who really want to find out. Described by Sean Latham as a 'reviled and disruptive literary form, thriving as it does on duplicity and an appetite for scandal', the *roman à clef* 'openly conceals fact within fiction, in order to mischievously muddle the distinction between them', with the 'key' to unlocking this distinction held by those readers already familiar with 'salacious gossip about a particular clique or coterie' (Latham 2009: 7). It resonates with the likes of *Popbitch*, originally a 'chatroom for music industry insiders to share their knowledge' that later became a popular gossip website and email bulletin (Birchall 2006: 93). In this vein, Niven positioned his character as a dark unconscious teller of uncomfortable truths: 'if I had heard it in private, then Stelfox was allowed to say it or think it', he claimed; 'the industry is a hugely racist and sexist institution' (Niven 2008). Even so, the book revels in the details of this world of chancers and sociopaths, unpleasantly enrolling the reader in its misanthropic glee. Moreover, for those 'knowledgeable insiders', *KYF* has additional appeal: many I spoke to purported to recognize colleagues (or themselves) within its pages, leading to games of 'guess who' in the office. Niven's autobiographical legitimacy was an authenticating badge by proxy – he was 'really there', and they could vouch for that because they were too. Unlike Stelfox, readers are fully aware of what is just around the corner. But ultimately, while the book is presented as a morality tale about a cynical world of greed, exploitation and complacency, as in the last days of Rome, its continued currency among an insider readership acts as a reminder of how little had changed.

Glamour, passion and cynicism

'Why do *X Factor* records sell?', mused Andrew, rhetorically: 'They sell because the TV format works. Is it music? Yes. Is it good music? Debatable.' Andrew, who described his own tastes in the last chapter as the 'underground subcultural side of dance music', diplomatically refrained from casting explicit judgement. Admitting that 'the nature of the recorded music business' is to be 'entirely dominated by selling', he is no Steven Stelfox. Nonetheless, the tension between what's 'good' and what 'works' repeatedly came up in internal conversations and aligns with contemporary press criticism. Some considered *X Factor* performances 'bland and derivative' soundalikes of already-popular acts; while others felt that the 'highly rationalized process of selection, construction and marketing' detracted from the romanticized ideal of artistic genius (Stahl 2013: 46). It also betrays an alternative reading of the *X Factor* as cynical emotional manipulation. A 2003 motion tabled by MPs condemned 'the irresponsible comments of the so-called music experts on this programme when passing their judgments on the young performers [. . .] pressurising young people, and women in particular, to conform to fit the mould' (Sheridan 2003). Through these pressures, academic critics were wont to consider such shows 'a mirror held up to late capitalist subjectivity' (Fisher 2014: 39). Such popular currency ensured that the show would be alluded to, unsolicited, by several of my interviewees as unfortunately representative – or more accurately, *misrepresentative* – of their industry.

Alert to public misconceptions, Comms manager Alan noted the distance between personalities inside the company and 'things like *X Factor* and Simon Cowell', the result of which was 'a false image that's projected around music'. Implicitly, his own depiction of colleagues – the 'guy in a Cure t-shirt'; the 'guy that puts a club night on' – presents the 'true image'. General Counsel Tony saw this as a failing: 'I don't think we've been as good as we probably could have been, in the past, about kind of getting across to the public actually just what happens in record labels, with a lot of really *passionate* people':

> *The music industry is a glamorous business. You've got things like the Brit Awards, people are walking down red carpets, whether they're artists or executives. [. . .] We're in show business, there sort of is a bit of glitz and glamour. [. . .] There is, I'm sure, a perception out there that the music industry is somewhat removed and detached from reality.*
>
> Tony. interview (2015)

'Glamorous' was a consistent reference in interviews – usually framed in opposition to 'passionate' and 'real'. Glamour is popularly viewed as an effect of commodification, fetishistically distracting us from the authenticity of the real thing. Semiotically, glamour is associated with 'selling', 'manipulation', 'seduction' – but it is also about 'making what is difficult appear easy', 'inviting just enough familiarity to engage the imagination, a glimpse of another life, utopia as a tactile presence' (Thrift 2008: 297–9). This tension is heightened in major labels since 'the glamor and intensity of the pop experience' is 'central to how pop music generates meaning, a focal point for individual and collective desire and a site for interpretation of authorial intent', as well as being economically crucial (Marshall 2012: 1). Interviewees were prone to elide distinctions, as Tony does, between managers and entertainers, picturing them both 'walking down red carpets' together – drawing out their resemblances as eccentrics with big personalities. Among staff with more mundane roles, the misconceptions were felt viscerally: 'When I tell them what I do', lamented product manager Ash, people 'always think there's loads of glamorous elements there, and there aren't really'. This frustration revolved around a sense that their jobs weren't viewed as 'real' work. Poppy reinforced this point by denouncing those job applicants superficially attracted to industry 'bling' – seeing only gigs, perks and backstage passes – drawing parallels between job-hunting and conspicuous consumption.[4] Veteran executive Leonard drove the point home with his diagnosis of a broader cultural malaise: this was an '*X Factor* generation', pursuing overnight success in creative careers simply because they 'don't wanna go to school'; that is, pay their dues and do the graft. For both, these aspiring employees appear as mass consumers deceived by the allure of a glamorous industry, missing that passionate commitment.

The talking cure

In different ways, both *KYF* and *X Factor* work on the premise that the industry's internal world is of considerable interest to a wider public, emphasizing the entwinement of commercial decision-making (which artists to sign, invest in and promote, and which to listen to) with the more affective and sensorial qualities of glamour and passion. Both are also 'texts' that circulate within the industry itself, often taking a critical inflection. These observations begin to draw us inside Sonaverse, where, I argue, a similar dual dynamic played out. Alongside technical standardization, regulatory change and cross-media marketing strategies, the pursuit of a new stability also involved a series of critical encounters being staged within the organization. From the mid-2000s, two business functions gained increasing centrality within the major label: consumer insight and corporate communications. Their position straddling its input and output boundaries became increasingly important, as an anecdote from Insights director Debbie makes clear. She recalled a moment towards the end of the Tony Blair ministry. A political advisor had been consulted by the CEO, leading to a 'considered' and 'strategic decision' to develop a joined-up set of initiatives across consumer insight, corporate social responsibility (CSR) and political lobbying:

> *We were working with [X]. So he was an advisor to [UK Government], a kind of freelance strategy advisor. And I don't know, I couldn't say who was responsible for what, but he was working closely with [CEO] and [Head of Comms]. That coincided with this [consumer research initiative] – being much more transparent about what we did as an industry [. . .] much more vocal – and that was the [CSR initiative] thing, it was all around that time – [and] reaching out to government and lobbying. [. . .] I was delighted to be a part of it because . . . a lot of the [consumer] research informed that. I'd like to say it was the research that started it and I feel like it was. But you don't quite know, I don't know what [X did] . . . I don't quite know . . . He was really interesting. He was like a shadow. Shadowy.*
>
> Debbie, interview (2015)

Debbie describes her own consumer research role working in sympathy with the 'more vocal' role of Comms, effectively becoming the company's 'ears' and 'mouth'. Crucially, this dual function was deployed not in relation to specific acts or releases but to Sonaverse itself – even to industry as a whole – underlining its legitimacy to represent them. Yet there is an interesting tension between the sense of increasing 'transparency' and of such decisions being made among 'shadowy' elite circles. Debbie characterizes political influence as a trade-off between light and dark: the dalliance with government represents a deal with the devil, in pursuit of an ethical imperative to protect the rights of creators and businesses, while becoming more accountable to consumers.

As Debbie suggests, the discourse of CSR became prominent for music organizations and their conglomerate owners around the same time. Vivendi's CSR statements, for instance, justified and pursued commitments 'to protect minors, to promote cultural diversity, to raise awareness regarding sustainable development, to foster access to ICTs and to share knowledge' – even if predominantly, according to communications scholar Marisol Sandoval (2014: 194), in support of financial performance. Externally, this might involve outreach work with schools and universities, charities such as the Nordoff-Robbins organization for music therapy or environmental group Julie's Bicycle or turning inwards to make a 'business case for diversity' in the workforce (UK Music n.d.). Accordingly, Alan oversaw a communications package of such CSR initiatives, alongside the standard PR repertoire of handling press queries and monitoring brand reputation in the media. He saw a need for the company to forge its own 'voice', separate from its artists'. Their statements, which were 'always very considered', he said, sought to reposition the company with a coherent and powerful self-interpretation that could combat public 'misconception'. Through this, he argued for the necessity of communications in the face of cynical interpretations of Sonaverse's place within the industry: 'first and foremost it's reminding people that we are not scary corporate monsters'. In this respect, he described a responsibility to both promote emerging social media tools among staff members and to encourage 'best practices' – that is, using them with care.

The increasing prominence of CSR is matched by the emergence of consumer insight – a subset of what would formerly have been termed 'market research' that emphasizes qualitative aspects such as reasoning, tone and feeling – around the same time. The name signals a form of professional expertise which, distinct from traditional marketing, seeks to 'mobilise the consumer' within the organization, as cultural sociologist Tomas Ariztia (2013) puts it, defining and qualifying them in new ways to clients. As Head of Insight at Sonaverse, Jeff distanced his work from the 'sadly quite accurate negative preconceptions' associated with 'the traditional world of market research', assembling customer databases and focus groups. He acknowledged continuity: it remains important to identify target audiences, asking 'what are they into; how do we get to them?' But insight activates a more holistic and humanistic understanding of what audiences find meaningful about music, often located within broader discourses of, for example, gender or nation. It is about presenting 'an inspirational and specific "truth" about the consumer' that goes qualitatively 'deeper' than statistical data (Ariztia 2013: 158). Jeff's department was not attached to a particular sublabel but to a central organizational unit run 'like a little agency', servicing Sonaverse's different subdivisions as internal 'clients' within the building. He saw visual design as a fundamental component in the art of convincing others of the value of insights: 'We take the presentation of work very seriously [. . .] We never make horrible-looking, 1990s, Intel Inside, PowerPoint presentations. We just don't. The medium must match the message.'

Comms and Insight are not new business disciplines, they are merely updated names for the older press/public relations and market research departments – albeit refined with distinctive research methodologies or social technologies. But interviewees on both sides considered themselves to have adopted an educational role within their own organization, arguing for a more communicative approach – not without facing resistance. Sometimes this conflict fell along departmental (creative/non-creative) lines. More often and more forcefully, it was seen as generational. Jeff viewed the very existence of his department to symbolize an epochal shift:

> *It just speaks volumes, very positively, for how the industry has changed. In that, the fact that the [Insight] department exists at all, it is a manifestation of an industry that wants to not just talk to itself, it wants to genuinely talk to audiences about their opinions. I suppose it's about being less . . . Think about the myth of the record industry in the seventies and eighties, the myth is around almost being solipsistic.*
>
> Jeff, Interview (2014)

By aligning the 'solipsistic' tendency to a much earlier era, Jeff diplomatically declined to point fingers at colleagues. Digital marketer Poppy was less circumspect, however:

> *At least now what we have is we can speak straight to the customer, because for years and years that did not happen. It's still a mindset shift for a lot of senior management, they think they can push a TV ad on to people and that's what's gonna sell. We think there's also value in looking at insight, consumer insight, and analytics to try and figure out how we should be doing things better.*
>
> Poppy, Interview (2015)

Equally, for Alan, digital disruption was a major challenge for:

> *a generation that were used to: make album – promote album – sell CD. And that*
> *was quite simple. For, you know, my generation – my direct peers at work, people*
> *of my age – because we've grown up with the technology, and we're in a technology*
> *company effectively, we've adapted and survived in the digital sphere.*
>
> Alan, Interview (2014)

Broadly speaking, then, Comms and Insight also worked together to implement an internal 'mindset shift', largely among an older generation of executives thought to have internalized a hypodermic model of 'pushing' products onto passive consumers. The younger generation saw themselves embodying an ethos of speaking and listening to active audiences.

Debbie described a particularly dramatic example of this encounter. She considered her major intervention to have been reframing the internal conversation on 'piracy' through a company conference involving talks from key practitioners, academics and thought leaders. More provocatively, she attempted to 'challenge people's assumptions' and 'get them to think' by recruiting a number of teenagers who 'agreed with the statement, "I never pay for my music"'. After a series of talks:

> *We kind of pulled back this curtain – I think, almost literally – and here was these*
> *kids: you know, the enemy! [. . .] They were a great set of kids [. . .] they were just,*
> *like, 'why should I pay to make your fat cat pockets deeper? Why should I? I've got*
> *no money and you're asking me to pay ten pounds for a bit of music?' And it was*
> *really interesting to hear A&R people getting really angry back! And at one point,*
> *[laughing] I think they asked [the CEO] how much he earned – I mean it was really*
> *fiery! [And he said] 'Why should music be free? It's made and produced and crafted*
> *with as much effort as a car; you wouldn't expect to get a car for free so why do you*
> *expect . . .' It was really impassioned and eloquent. It was just a really great debate.*
> *We were all talking about it for months afterwards.*
>
> Debbie, interview (2015)

In Debbie's description, then, Insight is ideally a process of listening that provokes discussion and re-interpretation. It achieves this by conjuring a pre-figurative microcosm of the industry as a whole, characterized by a reflexive process of 'impassioned' argument and 'debate', continuing well after the event, about the 'craft' of producing and selling music. There is a literal unveiling of the 'real' consumers, bringing their critiques inside the building walls to disrupt the broken model. Debbie's account paints the exercise as a therapeutic intervention: executives in a state of denial are confronted by a traumatic reality (of 'empowered consumers') and, through interrogation of difficult heartfelt subjects, are stirred into long-term self-reflection, externalizing their thoughts as they slowly move towards coming to terms with the problem. Insight as talking cure.

Critical encounters

What does the consumer *really* want? This quasi-psychoanalytic question led to a search for re-evaluation within industry and a professed ethic of atonement for past sins. What needed to happen now was to embed the process of continual active listening within the organization. According to the economic historian Albert O. Hirschman, the case for capitalism in the seventeenth and eighteenth centuries was lent a powerful moral force by its apparent ability to tame the dangerous allures of 'the passions'. By organizing economic life around the collective 'interests' of commercial incentives, rather than rational planning or the selfish pursuit of personal pleasure, he writes, classical liberals considered the market to reconcile 'the better nature of each, as the passion of self-love upgraded and constrained by reason, and as reason given direction and force by that passion' (Hirschman 1997: 43). Plato's two horses were to be kept in check by market discipline and the profit motive. Niven's A&R chancer Steven Stelfox has little interest in music; his passion is a nihilistically rapacious, ego-driven death drive given calculative form by the rampant commercial interests of 1990s musical capitalism. He would doubtless view *X Factor* only through the lens of Q4 sales figures. Whatever their merits (or otherwise), such texts are primarily of interest here for the responses they provoked among the corporate workforce. While my interviewees were typically avid consumers of industry mythology, happy to revel in the outlandish exploits of a *Kill Your Friends* for example, they looked askance at what they saw as commercial cynicism: staging performers' authentic lives and voices for viewing figures, while masking the real passions of those working to support them.

While appealing, this framing reproduces a common but unhelpful dualism around ideas of human emotion and passion, as described by the theorist Sara Ahmed (2014: 9–10): an 'inside out' perspective, in which emotional responses are affirmations of our inner authentic selves (we 'let it out'); and an 'outside in' perspective, where we are blinded and deceived by our emotions (we are 'taken in'). This chapter highlights two organizational examples of the critical encounter: the Insight department, listening out for shifts in audience feeling; the Comms team, busily producing narratives and images to counter the views of the cynics. The imagined active audience thereby became a more forceful presence inside Sonaverse, just as it did on-screen and in fiction. Both were engaged in rendering the *culture that produces industry* as an object of public consumption and internal reflection, at a time when its stock was especially low. But as we step into the office, we should be wary of this dualist emotional politics. Instead, we will need to listen with care to those critical encounters, and the passions they stir, wherever they may arise.

Passion work

Workers at the gates

Maybe it's just having worked amongst them for some time now but it's super-easy to spot the company crowd amongst the other passengers on the train: the young guy in the double-denim jeans and jacket combination, indoor sunglasses; another one reading the Steve Jobs biography; the two girls both in skinny jeans and loose-cotton vintage tops with glowing tans, discussing their festival plans. I get off at my station and sure enough they join me in the throng as I head towards High Street Ken. I am suddenly very aware of the sheer number of denim shirts I'm jostling alongside (denim is in). I catch snippets of conversation. 'Deep house and techno sort of stuff, yeah it was great'. 'Rizzle Kicks and Katy Perry . . .' Emphatic descriptions of last night and that guy. 'Do you want a Berocca?'

Inside the office, bacon sandwiches and bottles of Coca Cola are devoured by two hung-over colleagues with heavy, dark bags under their eyes. I switch on my computer and edge past a girl applying mascara at her desk to the kitchen for my cereal. Giggles and heavy swearing. Some drum 'n' bass suddenly blasts out. Back at my desk I open Microsoft Outlook: the CEO has sent a company-wide email about opening up a new area of business. It's full of rhetoric about 'entrepreneurial spirit', 'artistry and musical diversity' from a 'world leading music company', the need to 'bring the best music to fans around the world' and 'develop the great label executives and innovators of tomorrow'. There's some mild acknowledgement around the office – and a slight but definite tinge of bitterness. 'I can't read this, I'm going to have to delete it'. Next in the inbox is Record of the Day – a music news aggregation list to which we are all automatically subscribed. I languidly scan through its bullet points, while the office goes quiet around me. We settle into the rhythm of the working day.

Research Diary (2013)

This extract from my field notes documents a group of office workers assembling and slotting into the roles and activities that effect their working identities. I am far from the first to train my attention on the threshold of the workplace in this way (Farocki 2001). In his pursuit of the secret of capitalist profit-making, Marx led his readers beneath the market's 'noisy sphere' and past the sign that reads 'no admittance except on business'; experimenting with early cinema techniques, the Lumière Brothers trained their camera lenses on workers at the factory gates. A building's entrance marks the point at which an occupational community comes into being and disperses again, dividing the regulated time of work from the unregulated time of leisure – as thematized in so many clock-watching chart hits over the decades: five o'clock world; Friday on my mind; livin' for the weekend, and so on. Prior to taking up their place within the organizational division of labour, it is possible to spot qualitative differences between my research participants and many other white-collar employees. What sets the former apart is, to use the language that makes most sense to them, their 'passion for music'. Some evidence for this can be observed, some heard and some merely felt. Clearly, we will need all our senses about us as we move into what Marx described as 'the hidden abode of production'.

And yet, by contrast, once inside the building, the signifiers are eerily familiar. Supposedly, as the cliché has it, 'no two days are the same' in this job, such that value is derived from spontaneous expressions of individuality and unplanned interaction. But my diary extract conveys something rather more cyclical: the commute, the morning switch-on, the rhythms of work, the listless email scroll and the everyday social rituals. The aim of this chapter then is, first, to describe these habits, routines, sounds and styles, which circulate and make up the cultural life of the office in ways that are both spontaneous and tightly managed. Second, I want to make sense of them within a broader shift of market imaginary: 'passives' and 'actives'; 'pirates' and 'true fans'; and associated narratives of 'conversion'. Finally, I consider how the ubiquitous notion of the 'passion for music' manifests within these contexts, lending work meaning and prompting tensions that must be – but often fail to be – managed.

Anything goes

Let's return to the threshold of the building. The entrances to corporate headquarters typically confront visitors with an explicit commitment to culture. An impressive, artfully-lit foyer showcases music videos of artists across multiple screens. Through the swipe-card gates, a hallway boasts a spectacular wall-to-wall display of gold discs. An infographic mural runs up and down the surrounding walls of a central atrium, reeling off artists and key events by decade: '1970: Michael Jackson, Ray Parker Jr, ELO, Sony invents the Sony Walkman; 1980: Philips invents Compact Disc Digital Audio, Adam and the Ants, D-Train, MTV launches. . .'[1] Dominating an airy glass-covered space, one wall has been tiled white to recreate Pink Floyd's album cover for *The Wall* and another depicts a stencilled Led Zeppelin airship crashing through its red bricks. These carefully curated internal displays and ornaments blur company and artist histories. Evading the more difficult moments of mergers and buyouts, they reframe the corporation as a cultural institution and narrate a sense of pop music heritage, at the forefront of which today's employees can imagine themselves. Further inside, the offices themselves are invariably open-plan, deliberately fostering an 'organic' cultural flow between creative and administrative employees – and the occasional visiting star – through a design choice that symbolically rejects bureaucratic hierarchy. Here again, walls are adorned with screens, murals and speakers, while workspaces are fringed with posters, photos, toys and crafts, alongside files and product schedules. The most lavish or distinctive desks seem to be those of administrative workers (it is tempting to speculate, in a psychoanalytic mood, that these acts of decoration sublimate the frustrating absence of creative fulfilment elsewhere in the job). In any case, it is an acknowledged means of managing personal space and work identities (McCarthy 2004).

A more obvious mode of expression is fashion; this is the archetypal 'no-collar' workplace after all (Ross 2003). As a member of HR put it to me:

Smart casual with your own personality is the order of the day. Pretty much anything goes, I've seen it all. For your first day, maybe jeans and a shirt/t-shirt. (pre-employment email)

While 'smart casual', 'personality' and 'anything goes' are intended to imply freedom for some, they also suggest a level of complex nuance. Employees were certainly trend-aware, and here the timeworn division between 'suits' and 'creatives' (Nixon and Crewe 2004) is not one of role but seniority: only the most senior executives could justifiably be counted among the former, whereas sartorial expression is as common in the back office as the front. The suit, as Nixon (2003: 146) notes in the context of advertising, connotes 'corporate loyalty' over 'individual identity', helping to clearly 'demarcate the domains of work and leisure', a split that *creative* work resists. As such, the business suit was the domain of the 'C suite', where it displays status and power, beneath which it is rare – unless a designer tweed or vintage Paul Smith, perhaps accessorized with a fabric tie or artful pocket square. That is, it is a deliberate expression of personal style rather than a signifier of conformity. Alternatively, a fitted blazer might be worn with an open-neck shirt (or v-neck tee) tucked into high-quality jeans with patent shoes: an increasingly visible ensemble associated with entrepreneurship and the world of tech – but no less 'creative' for being so.

The office stereo presents further opportunities to display taste, style and enthusiasm, demonstrate insider status and perform camaraderie. My interview consent forms contained a request to allow 'written description of the immediate environment': reading this, one senior interviewee chuckled quietly and muttered, 'yeah, like: everyone was playing music'. Laced with knowing irony – at that moment, the office was silent – the comment showed a reflexive managerial sensitivity towards facets of work-life that might be considered ethnographically distinctive. Likewise, co-workers were highly aware of the performativity of office music: putting some music on is not organic, neutral or unconsidered. Each member of my own team had idiosyncratic tastes, a desire to discover new music – and a will to 'educate' others. There were clear hierarchies and boundaries of acceptability and taste. Often the soundscape became a site of territorial conflict: 'Oh God, no more hipster rubbish'; 'What, you've never heard this before? *Wow*'; 'This is awful, it's just noise'; 'Wait – do you *genuinely* like Oasis?' Open-plan design obliges personal music selections to jostle alongside promotional audio on loop – *THIS!* . . . *is Massive Anthems! 40 of the BIGGEST hits.* . . *!*[2] – in raucous competition. 'It can get bloody loud and really annoying' admitted product manager Rich; 'when you want to focus, it's a pain'. Yet he also saw himself learning through aural osmosis, keeping up with 'what's going on in the business' – what others are working on beyond his own team.

It was common, encouraged even, to be away from the desk, to chat, make jokes, blast music, play tricks on one another and share worries over a coffee. At times, the office buzzes with such interaction and exchange – convivial, spontaneous, sometimes unruly, textured with competing tastes and styles, and configured as a space of play (Ashton and Giddings 2018). Music helps colleagues to socialize across departments and to collapse work-life boundaries as they transition beyond company walls, into the more leisurely space-time of 'Friday-night drinks' (Gregg 2010), or even Tuesday-night drinks – perhaps via the office bar: 'we have drinks in the cafe, they have a bar open on Thursdays and Fridays', noted Data Manager Nathan; 'the bosses understand that there is a need to get us together'. These provide opportunities for networking, off-the-record information sharing and the forging of professional relationships – as well as displays of excess that accord with industry mythology: traces of pubs, gigs and clubs evident most mornings in the tired bodies, shared stories and pharmaceutical vocabularies.

In such ways, an informal domesticity pervaded working life, giving rise to loose-knit familial structures that ignore job function. Unkempt desks became home-from-home: a place to have cereal, apply make-up and recover from the night before. The vitamin-supplement Berocca, fizzing fluorescent yellow in a glass of water, was a common indicator of a hangover needing to be cured. There is a performative dimension here too, of course (a lifestyle being visibly lived), but also a social one: the offer of Berocca to a suffering party with a wry comment ('quiet night in, was it?!') might become a bonding ritual – a common experience, a shared sympathy – again crossing institutional divides (front office/back office; junior/senior). It can provoke discussion of musical passions and the recounting of shameful anecdotes across generations. The *professional* blends with the *confessional*: connections with workmates are formed after seeing one another 'at their worst'. Yet sometimes the disconnect between the past and present is palpable. One older executive, schooled in club culture, pondered a generational shift over lunch: compared to the haze of the nineties, all he saw were 'lattes, Graze boxes and boot-camp fitness training' – signs of a more 'mindful labour' (Gregg 2018).[3] Yoga, massage and well-being sessions were also possible on company time, helping employees become fitter, happier and more productive, as Radiohead forecasted fifteen years earlier. This dialectic of well-being and excess suggests a kind of hedonic self-management that differs from the utopian vision of weekend release formed in the previous century. Perhaps this accounts for its lack of visibility to the bemused Gen X executives: it had become simply an unexceptional moment in the extended routines of work sociality.

The office on display

The performance of corporate music culture is far more visible internally – through walls, desks, clothing, speakers and social rituals – than through any public-facing façade. This is the culture communicating to itself, after all. Equally, however, in line with a heightened communications function (as explored in the previous chapter), this culture also became distilled for external audiences – potential clients and employees – through a revamped online presence: more detailed websites and active social media accounts. Some of this content clarifies corporate structure and history, making much of the heritage of various sub-departments littered with imagery that remediates the internal aesthetics. Interestingly, however, much of it foregrounds the workforce: regularly profiling a 'featured intern' as part of a recruitment section, for instance, or hosting 'day in the life' interviews with employees, replete with personal music tastes, recommendations and handpicked playlists. On social media, the line between inside and outside could become very blurry, sometimes but not always deliberately, which increasingly needed to be managed during this period. Corporate accounts experimented with an informal tone, interacting with individuals. For a short while, Sonaverse ran a dedicated Twitter account, typically handled by an intern, taking followers 'behind the scenes' of work in the building. Meanwhile, employees placed their company affiliation in their account profiles to identify themselves as insiders – perhaps to be quoted in the pages of *Music Week* as a reliable source of industry gossip. Both personal and organizational accounts shared images of staff and artists around the office, quoted employee opinions, or shared and solicited playlist recommendations. Over time, this was ironed out a little as Comms took a more active role in monitoring public commentary:

> *If you go on [Twitter] and you tweet, 'x artist is a dickhead and I'm sick of them' . . . You just shouldn't do that. It's unprofessional. [. . .] We've not had as much of a problem with it in the last twelve months as we did previously: [. . .] the company has developed digitally and the understanding of why we do these things on the internet, and what the impact is, has grown.*
>
> Alan, interview (2014)

By 'just discouraging people from being idiots', Alan saw a need to distil and display a sense of professionalism as a public ethos, binding individuals, organizations and industries in a coherent identity.

The will to display a cultured, passionate and fashionable workforce took other forms too. It led to spots in *Elle* magazine's monthly 'Work Spy' feature, for example, spotlighting employee dress habits within (media-adjacent) companies. Universal's London office was the location for a June 2014 shoot, and Sony two months later (Work Spy 2014a; 2014b). Across two-page spreads, staff members (digital campaign manager, legal assistant, senior marketing manager, junior radio plugger, product manager, PA, label manager) are photographed around the workplace and presented, as in any such shoot, alongside details of the brands being sported. They match items from high-street stores with independent designers, band merchandise, and repurposed items. In embedded quotes, they self-style as 'romantic with a rock edge', with a penchant for 'vintage dresses that remind me of 1960s soul', or closely influenced by the artists they work with. One considers his hair 'pretty out there; I change it every couple of months'; another owns 'a treasured archive of vintage band shirts'. Comfort is important for a 'life running from the office to gigs' – luckily, 'working here means you can experiment with casual clothes'. For marketers, artist relations, administrators and legal executives alike, some formation of 'smart casual with your own personality' is indeed 'the order of the day' – not a particular genre or style but an actively thoughtful and confident dress performance. This aesthetic labour functions in the gap between self-expression and professional obligation (Entwistle and Wissinger 2006; Petersson Macintyre 2014). Remediated in fashion magazines and online spaces alike, it becomes a resource for corporate identity, underscoring the role employees play in reproducing the company brand (cf. Carah 2014).

Fashion thereby becomes a means of constructing an internal community with a coherent sense of identity and communicating to the outside world. As a not particularly style-conscious newcomer to the industry, such expectations caused me some anxiety (hence the email from HR). My own comparatively 'scruffy' or 'mismatched' outfits were noted by colleagues – not vindictively but nonetheless pointedly – intimating an attendant mismatch with company culture. That another new entrant, freshly graduated, wore an ill-fitting suit on his first day became a perennially amusing anecdote; by contrast, a sharply dressed mid-forties executive carrying a skateboard was simply part of the furniture. Over and over again, although in very vague terms, interviewees made reference to the importance of cultural 'fit' – of being a 'certain type', as Tony put it. For instance, 'you have to be happy having three or four different types of music blaring when you're trying to draft and negotiate with another lawyer on the telephone', explained Sophie: 'if you're not that type of a person, you just wouldn't really fit'.

Working rhythms

Despite being a senior manager, Gavin confessed to a certain exhaustion born from his lack of fluency in a world that 'lives and breathes' music, as he put it.

After twenty years, he viewed the major label as a persistently 'cliquey' institution: 'promotions hung out with the cool people in promotion and marketing, A&R with various people'; and described 'feeling like an outsider within the business' because 'I wasn't *cool*, in the sense that I wasn't gonna fit in and I . . . didn't follow the fashion'. From a different perspective, as a younger employee coming up, Olive viewed the industry as a whole, beyond particular departments, in similar terms: 'you had to be a proper music geek to actually work in it. . . and I don't think I am'. Both Gavin and Olive resisted pressure to exhibit a certain kind of fandom – but their characterizations are subtly different. For Gavin, being 'cool' is about being 'current', 'up-to-date' or even 'ahead of the crowd'. For Olive, being 'a proper geek' is associated with a longer duration: 'going deep' into music, having historical awareness and so on. For some, a new haircut every few months is important; for others it is a stock of timeless vintage items. Different modes of engagement with music culture are implied in each case, with different associations of belonging and identity, but both convey a shared working subjectivity: failure to meet these expectations produced a sense of disdain and exclusion.

Entering the office is not, then, simply a matter of moving into a certain space but also a negotiation with multiple timelines. Just as the demand for productivity induces a set of techniques for managing one's time (Gregg 2018), music also intensifies temporal experiences (Tagg 1999). Both the manager's productivity strategy and the fan's immersion in music attempt to resist the imposition of 'clock-time'; both seek to surmount the common sense, quasi-'objective' narrative of technological progress compelling a constant linear forward movement: the determinist idea that one cannot hold back innovation or social and aesthetic change. As such, like others, they follow the sociologist Henri Lefebvre (2004), for whom the subjective experience of such change is ordered and patterned by multiple overlapping rhythms, cycles, repetitions, durations and flows, which structure the social production of space. Lefebvre was concerned with urban space, but similar observations can be made about the workplace. Equally, as Georgina Born has it, music mediates a 'multiplicity of time in cultural production', at different durations, from an immersed immediacy in unfolding sound through to long-term historical periodization, via a sense of stasis or innovation – produced, for instance, through repeated appeals to genre and style (Born 2015: 362).

Thinking about distinct musical temporalities, and their interleaving, is a generative way of understanding how individuals locate themselves in contexts of cultural change. For example, listening enters into the time of work itself and structures the working day, from the morning commute through to the evening gig. As such, music work complicates pop's long-standing appeal to lend meaning to (or, as Adorno would have it, to standardize) 'free time' and its repeated refrain of 'living for the weekend'. At a macro-level, the company makes much of its musical history through its displays of heritage and its robust institutionalization of the concept of 'catalogue' in organizational structure – meanwhile the past is re-evaluated (and revalued) through streaming technologies and licensing strategies, in the context of a seeming epochal break from the 'old' industry. All of this seemingly comes to a head in an algorithmic playlist culture that harnesses music to its functional capacities, generating income by regulating mood (Rekret 2019). What particularly interests me here, though, is how this plurality of music's temporal layers plays out in the polyrhythms of office life. The history of manual work songs, from agriculture to production lines, tells us something about how music mediates the 'rhythms of labour', enabling and constraining workers' capacity to generate experiences of concentration and distraction, build community and voice criticism – and negotiate the authority of superiors (Korczynski 2013). If the office building does not fully circumscribe the professional production culture of popular music, it nonetheless condenses and mediates that culture. Working within the music company intensifies these rhythms, but, insofar as employees are organized around producing music as marketable outputs, their professional 'musical time-sense' goes beyond this, 'aligning [. . .] work rhythms' within the office 'with the innovation and promotion rhythms of the commodity' (Cronin 2006: 618) – and, moreover, injecting it with purposeful intent.

To make things slightly more concrete, we can reflect on how day-to-day interdepartmental relations are typically structured around the 'release'. The very name suggests an elastic sense of time. Accordingly, the management of promotional campaigns and logistical lead times sets the pace of work, its intensity compressing and relaxing with the flow of products through the schedule. Different departments experience relative 'quiet' and 'busy' periods for releasing products – even if it is common to hear that there is *no* 'down time'. Product scheduling is partly about coordinating the flow of income across the business as a whole. Rich gave an example: over the summer, 'new releases dry up because it's festival season, people are out gigging [. . . but] you need stuff to sell in that time, so that's where the older artists, the heritage acts, come in'. This coordination activity structures the annual calendar. The slower (catalogue-driven) summer period is followed by 'Q4' (October–December): the most agitated and infamous months of the calendar, when pre-Christmas pressure is placed across the entire company to 'make the numbers' for the year as a whole. This involves a level of complex communication across divisions to avoid clashes between marketing campaigns or competition for resources. Secondary output (e.g. re-releases) must not obstruct frontline priorities, or they might play into a coordinated campaign of current and classic releases that includes branding, synchronization and playlist strategies.

The reporting of revenues each financial quarter of the calendar year provides an important fixing point for shareholders, steadying the sense that 'nobody knows' where the sucesses will be found (Caves 2000), and introducing a narrative of incremental, linear progression. But its cyclical nature also structures workers' temporal flow ('we're in Q3 already, how did that happen?'), while festivals, conferences, awards ceremonies, and Christmas parties all offer spaces of collision between staff and audiences (and employees' fan identities), plus opportunities for the industry to display itself to the wider world through television broadcast, social media or newspaper gossip column. But regular financial reporting is also used as a means of justification. In troubled periods, when poor financial performance may give reason to consider departmental restructuring or redundancies, the quarterly calendar may be watched with heightened anxiety. 'Frontline' labels, with their faster pace of change, are also associated with greater instability – for catalogue, 'it's not a daily pressure', according to Rich: 'I think if I was in frontline I'd be slightly more nervous'. In catalogue, additional dates like Valentine's Day, Mother's Day and Father's Day further pattern the year with periods of emotional and productive intensity, which, Leonard argued, bind these 'gifting markets' to the traditional physical album format: 'How do you add emotion into buying or gifting a digital product?', he worried, semi-rhetorically – suggesting that whoever answered this question would make a fortune. These periods remained the most important opportunity for physical sales, producing regular 'crunch' periods, with increased pressure on production to meet manufacturing and distribution deadlines.

This was not the only challenge faced. The traditional promotional 'feedback loop', in analyst Mark's words, wherein multiple weeks' radio play anticipates and follows sales charts, was no longer adequate to 'a world of infinite choice'; the challenge became to coordinate with the 'network effect' of online sharing, making use of streaming, branding and licensing to create promotional 'buzz'

– all income-generating activities in themselves. This is less about creating a pathway to a sale than enrolling anticipation in a particular release campaign. Staff will follow the marketing build of a high-profile project with fannish interest, adding to the anticipatory atmosphere. Beyond stimulating demand, though, the 'act' of consumption is located not in paying money but paying attention. As a result, concern over illegal downloads became increasingly displaced by pre-release 'leaks', which undermine the 'build' of a marketing campaign (cf. Lea 2015). But, as Debbie recalls, consumer research indicated that standard practice was considered incomprehensibly 'archaic' to a public who could not legitimately access the songs they were hearing. In response, a cross-industry policy – labelled On Air, On Sale – was trialled by the majors in 2011, ensuring that any track being broadcast on radio could also be downloaded and streamed (Lindvall 2011b). It was not a success. 'It changed everything', Debbie reflected, because 'there was no room to build a song'. Consequently, commercial radio playlists, assembled by projected and actual sales, took tracks with demonstrably *low* sales off air. Moreover, it required dramatic changes to internal 'just-in-time' promotional practice, reducing crunch times and demanding relationships that had yet to be put in place; the interdepartmental processes that were required to coordinate such campaigns – bringing multiple teams and departments together, sometimes internationally, linked with shared systems and regular planning meetings – were still in their relative infancy. The effect of the initiative was to render the importance of repetition painfully visible, for both constructing consumer tastes and regulating internal workflow. Such frictions caused the initiative to be rolled back later that same year.

Of pirates and passions, converts and cones

When I asked interviewees how their careers began, it was strikingly common to begin with a youthful account of being surrounded by music, or of their early love for certain artists. As our conversations progressed, they would then turn to other allegiances – but often revisit their younger selves in order to do so. Clearly, they located their career origins in fan-like attachments and, in so doing, sought to evoke a strong sense of continuity (Banks 2019). Towards the end of our wide-ranging conversation, for instance, corporate Comms exec Alan felt moved to defend and justify his career choice, or 'the point of working in music' as he put it:

I love what I do because . . . it . . . I love music. And actually now, you know, looking
back on three years, there's – I feel quite protective about the industry.

Alan, interview (2013)

Alan's wording is careful but faltering. As a communications executive, he was rhetorically skilled and, throughout the interview, I could feel him judging the implications of each conversational turn, in order to stay on message. While keen to emphasize how his co-workers were, first and foremost, genuine music fans, equally he knew first-hand the brand value this holds for the company: indeed, this is what he is paid for! But turning to his own commitments, his speech became disjointed, restarting and working over the same formulation. Perhaps, then, his role as company representative confronted him with an unresolved anxiety, disrupting his own communication in an attempt to avoid appearing empty and wooden. With much less confidence than he had previously projected, here he rapidly slips between three affective bonds: love for his job; love of music as a cultural object; and protectiveness towards the music industry. These subtle semantic shifts suggest a negotiation between multiple attachments as he feels his way through an imprecise set of feelings – a slippage that expresses the uneasy identification between individual and corporate identity.

Alan articulated his career biography as a struggle with the fan-pirate dialectic. While at school, he confessed, 'I was a rampant pirate'. Our similar age allowed us to connect through nostalgia for a former era of illegal downloading, leading to a lively exchange, trading memories of obsolete software and slow internet connections, in which he betrayed a certain ambivalence. In one moment, he described a conflict between music as a social and economic practice: 'I don't think I really considered it as theft [. . .] it wasn't really file-sharing, it was just like sharing with your mates'. In the next, however, he views it in a more cynical light: 'Fine, I'll take a copy'; or worse, 'I'll download their whole discography in one go'. With this, Alan reframes a legal argument as a moral critique of unthinking self-interest. 'It was just so throwaway. Do what you want, don't give a shit'. He was keen to emphasize the past tense, describing a former self-led astray by fannish desire into acting without due care and consideration. 'I had no clue. It was a lack of understanding about why piracy was wrong'.

I would like to attribute Alan's admission of this aspect of his past, despite his institutional position, to my skill as an interviewer. More likely, it suggests a broader change in the industry's health and a shift in its authoritarian approach to illegal downloading. The 'official' discourse, with which some individuals still aligned themselves, remained a simple sense of indignation: being 'passionate about piracy', as legal executive Melissa described herself, is being passionate 'about musicians being paid'. While the topic was not broached during our interview, for example, in a later serendipitous encounter with the more old school Leonard, he stopped me with 'one thing I do want to mention':

> *I do like the way in which they [the BPI] are treating the pirates and the torrent freaks and so on [. . .] They've got to keep fighting that fight. Hitting the government with that. I want to make that clear.*
>
> Leonard, interview (2014)

The pointed and deliberate manner in which Leonard, a veteran executive, wanted to register his support for trade bodies 'on record' is possibly best understood in the context of a media environment which distrusted them. It also speaks to the culture of taboos and discursive constraints on senior executives that former EMI finance manager Jonathan Wheeldon encountered in the mid-2000s, where, for instance, in high-level meetings, 'it felt personally risky even to consider whether file sharing might not be such a terrible thing, unless one had a PowerPoint deck illustrating how it could be brought under control and monetized' (Wheeldon 2014: 137). The prevailing orthodoxy a decade later, however, had moved away from the simple condemnation of wanton downloading, mirroring individuals' own responses. Describing discomfort with the term 'piracy' itself, Debbie distanced her own work from the 'scary messages' of trade bodies, which 'felt so alienating and unnecessary'. She argued:

> *You can't just call this bunch of people pirates and criminals. [. . .] You know, they're just the general population, we've got to consider them much more carefully. [. . .] I was much more interested in talking about legitimate music services.*
>
> Debbie, interview (2015)

Tony concurred, citing Spotify as an example of such legitimacy. Supporting new platforms was simply to 'make it [piracy] unattractive: you've gotta make it boring and we've gotta make the real deal just feel much better'. It is, of course, easier to talk about 'legitimate music services' when they are gaining in popularity. Yet the framework of ethical consumption around which such platforms were built was prefigured by the kinds of careful reflexive consideration of one's own passions described by Alan and colleagues.

Alan's biographical *mea culpa* mapped neatly onto, as he put it, 'the point of working in music'. It also echoes a long process of rethinking what 'the consumer' wanted, if it was no longer physical mass-produced products – and, moreover, who they actually were. A new cluster of organizing concepts sought to address this for a post-scarcity digital age, most notably Chris Anderson's (2008b) 'long tail' and Kevin Kelly's (2008) 'true fan'. Both former editors of the tech-culture magazine *Wired*, they argued respectively that the internet opened avenues to address a far more diverse range of consumers with small batches of niche products; and, because the most active of these invest disproportionate amounts of time, energy and money in the artists they love, cultivating them in manageable enough quantities – a mere one-thousand by Kelly's reckoning – could sustain a career. Liberation from old models might therefore be found in fan communities. There, emotional attachments could be leveraged into long-term commitments and even *prospective* economic support, through crowdfunding, for instance (Leyshon et al. 2016). The problem was not, therefore, that people didn't want to pay for music but that people wanted to pay often significant sums – but weren't able to. In Lovell's (2013) model, these 'superfans' valued ongoing social connection and personalized content; meanwhile, 'freeloaders' and 'gawkers' could come along for the ride since, less inclined to part with their cash, they could still facilitate word of mouth. The latter therefore constituted an internal audience for superfans' conspicuous consumption – and might later be tempted up the demand curve as paying superfans. Variable pricing for diverse content (or the same content in variable formats) solves the apparent problem of insatiable demand, 'empowering' fans with a multi-tiered set of pricing options that 'allows' them to display their particular commitment level. This suggested a particular moral economy in which affective investments led inexorably to financial investment – if you let them.

It was common to hear talk of superfans, long tails and crowds around the office or all-staff meetings, all part of the e-commerce *lingua franca*, alongside ideas about youth 'lifestyles', 'subcultures' and 'tribes' of the kind that, as explored in Chapter 3, accompanied the new countercultural generation as they transitioned into creative careers. More significantly, the underlying moral economy was operationalized through two marketing practices: *segmentation*, or market sizing and classification; and *conversion*, the quasi-missionary zeal to turn freeloaders into paying customers. In the first instance, standard demographic data – age, gender, nationality, ethnicity, socio-economic group and so on – were being displaced as the key predictors of consumer behaviour by 'post-demographic' notions of values and identities. Here, for example, is the 'FECI' fanBase model developed by Sony's marketers:[4]

This model is grounded in a granular lifestyle approach, foregrounding listeners as active audiences (first column): tribal 'users' of music for identity

Table 1 Sony fanBase model

Tribe/Lifestyle Categories (Examples)	Active-Passive Category	Total UK population (%)	Market Value (%)
Rage Against the Mainstream Mini Moshers Trendanistas The Curators All Tomorrow's Playlists	Fanatics	12	31
White-Collar Radicals #Hotrightnow Tipping Pointers Dad Rocks Domestic Goddesses	Enthusiasts	25	32
App Happy Pop Princesses Footie and Playstation Here Come the Girls Reassuringly Familiar	Casuals	28	23
Likely Dads Mum's Got Talent The Good Life	Indifferents	35	14

construction. This it maps against a ranking of engagement and intensity (second column). The origins of this lie in an influential piece of market research on digital music consumption from 2003. The portentously titled Project Phoenix, commissioned by Emap (a UK media and advertising company with a number of culture and lifestyle magazine titles in areas such as fashion, computing, games and music), suggested that audiences aged 15–39 could be divided into what they referred to as four 'passion zones': savants, enthusiasts, casuals and indifferents (Emap 2003). A similar model, developed for BBC Radio One in the mid-2000s, distributed listeners along a widening three-part 'cone' of 'scenesters' (a small number of tastemakers), 'restless' (a relatively large number of fans), and 'contented' (a mass audience who do not appreciate being challenged) (cf. Rogers 2013: 171–2). While Emap's segmentation was static, the cone suggested an image of listeners being drawn in towards the centre. Likewise, fanBase suggests a more dynamic movement, from passive to active. More specifically, it promised a powerful, fine-grained tool for listener *conversion*.

This responded to one of the more controversial disputes within industry circles: whether streaming's 'success' could be measured by the capacity of (unpaid, advertising-supported) freemium models to provoke casual listeners into upgrading as paying subscribers (e.g. Ingham 2015; Mulligan 2015a). The very notion that it might is suggested, again, by Emap's 'passion zones', particularly the version popularized by industry analyst David Jennings – who, echoing the popular psychology of Abraham Maslow's hierarchy of needs and motivations (from basic self-protection through to self-actualization), rendered it as a pyramid. Jennings' version was taken up among music streaming engineers, including for Spotify's Echo Nest recommendation system (Lamere 2014), helping them conceptualize listener engagement. In his ethnography of algorithmic recommendation, Nick Seaver (2022: 77–81) describes its use in such worlds in terms of a model of 'avidity', typically collapsed into a passive-active binary. The expressive tribal approach, constantly needing to be updated in line with the latest trends, was viewed with increasing scepticism, both within academia and the marketing profession (Gray 2007; Hesmondhalgh 2005). The avidity model, however, endured – perhaps because it offered an image in which engineers and executives could see themselves. The suggestion that 63 per cent of available value is located in the most active 37 per cent of the population would be an arresting statistic for any marketer looking to target their promotional spend, let alone one that's a self-identified avid fan.

And yet, as Seaver suggests, these frameworks also tell us something about how insiders 'see' the external socio-cultural world of taste, aesthetic discernment or critical capacity, allowing such qualities to be quantified, measured and, consequently, normalized as part of everyday corporate practice. As such, they promise a model of truth. While earlier definitions of 'passion zones' were derived from surveyed consumers' self-descriptions, streaming services generated more finely-grained behavioural data through which users' actions appear to reveal their *real* tastes and preferences, however they present themselves in public. 'Every time a listener adjusts the volume on the player', writes the programmer Paul Lamere (2014), 'every time they skip a song, every time they search for an artist, or whenever they abandon a listening session, they are telling us a little bit about their music taste'. Behavioural data analyses thus informed, and were informed by, software engineers' own self-understanding as 'passionate' music geeks. As Seaver (2022: 80) writes, in effect, the four-level pyramid reinscribed a long-standing fans/masses dichotomy: the 'lean-forward' users 'fiddling with settings, actively engaging with the interface, browsing, and skipping songs'; the 'lean-back' users 'looking to start the music and then leave it alone'.

This folk-anthropological insight mapped imagined listener archetypes against internal self-understanding, mediated and legitimated by the Maslowesque devices of the pyramid, or cone, and the 'purchase funnel': terminology used by marketers to describe and measure the 'consumer journey' from initial awareness of a product to eventual payment; or, from passive contentment (background radio, ad-funded streaming) to active engagement (tastemakers, paid subscribers). The notion of 'conversion', laden with missionary overtones, is crucial. Measures of passivity/activity were no neutral descriptors, they carried the moral weight of a crusade. The ethics of consumer avidity married perceived cultural value with economic value. This was important for the development of industry legitimacy insofar as it could underpin the reflexive moral project of work, in line with a broader 'turn to life' in corporate culture (Heelas 2002). In both informal discourse and formal metrics, practitioners could not just imagine and construct their audiences but also locate themselves. Parallels between fandom and religious faith are often overdrawn (McCloud 2003), and I hardly want to characterize record executives as repentant sinners, evangelically engaged in recruiting others onto the 'true' (fan) path of salvation. While resonances with this crude soteriology can be found in some interviewees' accounts, my argument is simpler: that the emergence of this relational and narrative-driven model of consumption is explained by its *ethical* weight rather than any technical capacity to somehow generate objective evidence.

Emotional dependency

You would read sleeve notes and you would know about, you would know that Atlantic [Records] were in Rockefeller Plaza in New York without having been there. You would know that this meant something, this, this resonated – and then you read about [Atlantic founder] Ahmet Ertegun and you learn about the history of Atlantic, and then . . . how that model, kind of, then applied. So I think you should, kind of, always have that, be able to make those connections, but you've gotta have that, kind of, thirst for the knowledge otherwise . . . otherwise I'd just get really desperately bored [. . .]. All I'm paid to do, is to read about the music industry, speak to people about the music industry, and write about the music industry. That's all I do [. . .]. Within academia, you'll have a particular subject . . . you go in very intensely and it's very passionate up to a point – and then you move on to the next thing. Whereas, I dunno, I'm just the, the, the kind of, childhood sweetheart. You're always kind of there and yeah you can nag at them and whatever else. And kind of, they wind you up and they kind of annoy you like no other. But you're constantly there, you can't . . . There's a kind of emotional dependency, just because you're so interested in it.

Ian, interview (2014)

As a trade journalist, Ian attributed his career and capacity to write to his obsessive music practices, starting with a younger version of himself immersed in his record collection leading to a long-term relationship.[5] Notions of love and passion are common in creative work and beyond. Consider the Spotify web developer, for instance, for whom joining the company was 'like finding a girlfriend, finally', his emotional relationship with work the culmination of a nomadic lifestyle of serial employment: 'It's like, "Yeah, yeah finally. She knows. She gets it. She gets it. Finally." I've been dating around, and I finally found my company. They get it' (Titlow 2014).

Interviewees did not always describe their employment contract in such questionably vivid terms but the metaphor of the relationship is an enduring one. The image Ian conjures is of himself as a young fan listening intently, on headphones perhaps, while poring over packaging ephemera, preoccupied by minor esoteric details. He considered these practices to have naturally foreshadowed his current position, exhibiting the movement that Alan described, between *passion for music, passion for work* and *passion for the industry*. This should prompt us to attend to an important characteristic of 'passion': that it is not an abstract quality, which one either does or does not have; passion is particular, intentional, produced and oriented towards and attached to some specific object (Bennett 2018a). In Ian's case, the object of attachment and identification slips – or rather, it is strewn with 'connections' – between the album's musical content, the details on its packaging, and into the processes of its commodification. The packaging (the 'sleeve') can thus be read as a text, for clues (the 'notes'): a name, an address, a building, populated by individuals, with histories, from whom lessons might still be learned – and so on . . . Ian experiences these logistical traces aesthetically, evidence not simply of a banal manufacturing process but of a cultural-industrial practice, one which also (in an apt metaphor) 'resonates'. These amateur inquiries into the commodity form have led him 'through the factory doors' and into encounters – figuratively at first, as an avid fan; but now for real, as a professional journalist – with, to use Marx's phrase, 'moneybags' himself: the executive record man. Here the passion for music becomes a method – a 'learning' strategy. But rather than revealing (as for Marx) 'the secret of profit-making', Ian learns how the record label 'meant something'. Restated using actor-network theorist Bruno

Latour's (2005) language, he expresses a desire to 'follow the actors' (just as fans follow artists perhaps), tracing the connections and opening up the 'black box' of their conditions of production. This leads him yet further, into normative territory – to outline what any industry participant 'should' look like: a fan with a 'thirst for knowledge' and insatiable curiosity. The repeated impersonal 'you' displays Ian's identification with an imagined community of similarly inquisitive listening and reading practices, one which valorises and validates the will to go 'beyond' the surface of the text. His own passion for music performs a kind of professional boundary-work (Gieryn 1983), setting expectations for inclusion in this community of practice: he did not choose music, he was always this way – as anyone seeking similar success must naturally be.

Clearly, Ian is an avid consumer and keen to portray himself as such. But is he an 'active'? The theorist of organizational discourse, François Cooren (2010: 59), tells us that emotion 'tends to be short-lived and sudden, while a passion tends to refer to a state, condition, disposition that somehow endures, lasts, persists, lives on'; equally, this is not a simple intentional choice, since the impassioned individual 'is moved, led, animated by her passion, a passion to which she cannot resist'. And so, Ian cautions against confusing the sustained long-term relationship of his 'emotional dependency' with the non-committal romances of the academic (i.e. me). Rather than commenting from a place of dispassionate distance, his work offers comfort, a place to be yourself, at home. This opposition between the childhood sweetheart and the series of flings is journalistically evocative, articulating the suggestive nuances of professional passion: youthful origins leading to a long-term labour of investment, with its 'ups and downs'. It also resists the dualist inside-out/outside-in model of passion that Sara Ahmed cautioned against in the previous chapter. Rather, according to Antoine Hennion (2007), passion is a term that describes reflexivity towards one's own attachments. The paradigmatic passionate figure, he suggests, is the amateur enthusiast – with 'amateur' understood not as 'unprofessional', absent of skill, but more etymologically: 'someone who loves'. Rather than self-mastery and autonomy, this is the *action of being acted upon*, of allowing oneself to be animated by music. Passionate workers are thus 'turned towards their object in a perplexed mode . . . on the lookout for what it does to them, attentive to traces of what it does to others' (Hennion 2007: 104). The recorded music commodity acts as a boundary-crossing object for them. It does not arrive sealed-off, inert, ready to be 'consumed', but as an object whose meanings can be disassembled, navigated and reconstructed in situated encounters. To have a passion for music, as my interviewees describe it, is not simply to be 'active': it involves reconceiving agency as an act of submission, giving oneself over to an object or others. It is to be attentive to music's own agency, how it draws listeners into a rich semiotic and affective world, encouraging them to travel with curiosity between moments of its creation, reproduction and reception.

Realistic passion

Music pervades everyday office conversation, through reminiscences of gigs attended and in comparisons of songs and classic albums, through which tastes may be policed – gently, ironically or more vigorously. Even so, it is not immediately clear why a 'passion for music', often deeply felt and richly articulated, is so often seen as a precondition for working in this context. After all, most employees' personal convictions have only minimal bearing on the character or output of their job function. Perceptions of quality remain important: for Rich, 'at the end of the day, as music fans in the industry [. . .] what we want [to make] is a great quality product' – whether in design and packaging choices, 'finding the best audio' or 'the best article written in a book that goes in the packaging': put simply, 'it's nice to see people do still care'. Nonetheless, interviewees typically marked a sharp distinction between their identities as consumers and professionals. Expressing this most succinctly, Tony argued that, while Sonaverse employees tend to conform to 'a certain type', one needs to be able to bracket out particular preferences, where they do not coincide with the object of work:

> I think if you were only prepared to work on music that you liked you wouldn't have a job, you've got to do that. [. . .] It helps, I suppose, if you like the artist but ultimately you've gotta be able to work with an act whether you like the artist or not.
>
> Tony, interview (2015)

Tony's formulation implies that the two are usually mutually exclusive; there is no room for purism. Offering a more elaborate explanation, Rich described the disjuncture between his tastes and his job:

> I love mathrock. I absolutely love mathrock. [. . .] But if you want a career in music [. . .] you've got to be able to see the bigger picture. If I only wanted to listen to mathrock and work on mathrock I'd be living in a tent under a bridge somewhere.

I wouldn't . . . You know, I've got to be slightly more career-focused and realistic. So there is matching the romance of it with the real world expectation.

Rich, interview (2015)

Here, fandom appears as a youthful romanticism indulged in by those ignorant of 'the bigger picture'. This distance is 'something that I've become more comfortable with as I've aged' – since, for him as much as Tony, the alternative to being 'realistic' is being unemployed. Commonly, interviewees drew on strict art-commerce distinctions in order to exhibit their passionate allegiance to the former but realistic commitment to the latter. 'If you work for an indie, it's maybe slightly different', Nathan suggested, 'but working for a major, you're going for the dollar, and sometimes [laughing] the best tracks aren't the best tracks!' Ash agreed, offering a sanguine inversion of normative assumptions, by stating his preference for commercial success over aesthetically-worthy projects: 'people can talk all they want [. . .] but if you've made loads of money and sold loads of records, that is essentially what we're here to do'. His quiet sense of vindication derives less from having 'made loads of money' than from defying a critical consensus and puncturing the elitist judgement of colleagues' 'talk'. As such, Ash agonized far less over a sense of a job done well for its own sake than those appealing to 'realism' as a justification. Responses ranged from resignation to cynicism to retribution, but they were clear that 'commerce' typically wins over 'art'.

Like the recommendation algorithm engineers, label staff are caught between competing 'evaluative schemes' (Seaver 2022: 82): their personal tastes and preferences and a larger professional sense of what makes music circulate widely and effectively. And yet, personal passion does not entirely disappear in service of the common good. Interviewees commonly downplayed *specific* tastes but talked up their general avidity as *omnivores*. This became a central element in making their own work engaging and meaningful, positioning their jobs in terms of building frameworks that enable *others'* enjoyment. While Rich performed his own active fandom, confessing an *intensity* of feeling for a *niche* genre, he described being hugely 'aware of the heritage' of the company, for example, and gently chided colleagues on the front line whose historical understanding he suspected was lacking. Far from abandoning his love of music, then, he *puts it to work*. This points to a messier relationship between the affects of consumption and production than a simple art-commerce tension. Beyond personal aesthetic preference, discussions commonly revolved around the hybrid cultural-technical trade knowledge of 'what makes music popular': the crafts of marketing, managing and negotiating over music might be just as – perhaps far more –

interesting than those of songwriting and production; one might appreciate a 'clever campaign' or a 'great deal' as much as a 'memorable hook' or a 'soulful voice'. In the interpersonal communication of work – networking, selling one's attributes and acumen – employees mobilize their tastes and opinions as part of an appreciation of the whole process of industrial music production. This could be both utilitarian and genuinely meaningful.

As a new entrant into the workplace, confident in my own tastes but unused to finding common ground, I was initially disconcerted by the apparent superficiality of passing conversations – half-personal, half-instrumental – I was hearing in lifts, in the café, on the way to the kitchen, in the pub:

> *Just heard the single by [New Young Hope], was that one of yours?*
>
> *No, that was [Product Manager], great track but a nightmare to work, apparently. One of them jumped out a hotel window and broke his arm.*
>
> *No way! Sounds like [Next Bright Thing]. I've been working on their video.*
>
> *Oh really? You should speak to . . .[6]*

Such a mode of discourse mixes multiple genres of talk, from relational forms of phatic communication, small-talk and gossip to more transactional and goal-oriented forms of enquiry and planning (Koester 2006). The fluency of this rapid genre-switching hides the skill of its construction. It performs the interpenetration of work and life in a manner with which I was naïvely unfamiliar, accounting for its peculiarity to my ear. This is not just incidental – it it is how a sense of 'fit' manifests within the workplace as a communicative skill. One's mastery of this kind of talk appeared to correlate with career success; the highest-performing individuals can be recognized through their rhetorical virtuosity. Those who prefer to retain a strict work-life/art-commerce balance may find their ability to ascend the ranks rather less rapid. It may induce anxiety: a feeling of being out of place or not belonging, of being an 'outsider within the business' (Gavin). Simply put, both 'getting in' and 'getting on' are the result of constantly realigning one's identity with the imagined community of the broader industry.

The feel of work

Things did not always go so smoothly, as a final vignette illustrates. By 2011, Sonaverse was redirecting resources by making its back catalogue central to revenue generation. The catalogue label changed management, headed by a frontline veteran, who implemented a new office layout and cross-departmental workflow, integrating admin and marketing teams. The mixed modalities of frontline and catalogue began to intermingle: the pace of work accelerated, the (older) 'geeks' saw the injection of a (younger) 'cool' crowd. Mindful, perhaps, of the image of a dispassionate office just like any other, redecoration swiftly began (an artists and album-sleeves theme). However, it soon became apparent that the label was still not considered 'music industry' enough; it did not communicate the appropriate impression to potential artists, clients and guests who might circulate on the floor. A directive filtered down through the management hierarchy: music must be heard at all times. Various individuals on the floor, considering themselves particularly active listeners, were bemused. The arrival of Spotify in 2008 had produced a socializing effect on my own team – particularly its collaborative playlist function, where multiple contributors curate a shared soundtrack. This presented opportunities to experiment. We frequently challenged ourselves with ever more obscure playlist themes – Guilty Pleasures; Metal Monday; Best of '00s; Underrated Duets; Songs About Birds; I Need a Doctor; I Bet You Didn't Know That Sample; Nice Voice, Shame About the Face – each selection being broadcast to the rest of the floor. We would push the boundaries of each other's tastes, mischievously inserting Celine Dion or Simply Red in the midst of an earnest 1960s garage rock playlist. In such ways, the collaborative playlist had become a fixture of working life. Yet the directive was reiterated personally several times, increasingly sternly, on quiet afternoons. True, silence was indeed sometimes preferable and personal headphones were an essential, if somewhat antisocial, technology in order to 'zone out' from the hubbub and focus. Still, the managerial injunction was perplexing. In a more insubordinate mood, we placed bets over email on just how much 180-bpm jungle or free jazz was required to make a senior executive slam their door shut – until eventually threatened (quite seriously) with an enforced diet of BBC Radio One if we did not listen 'responsibly'. In a final riposte, my colleagues contrived a plan to maintain a constant audible soundtrack: placing a single CD compilation album (a deliberate choice of 'old school' format) on permanent repeat. Meanwhile, they continued to bury themselves in headphones. The directive was quietly abandoned shortly after. Order – or the lack of it – was restored.

Music's function for workers has long been ambiguous: not just facilitating community and distracting from boredom but also enhancing productivity and silencing dissent – while occasionally enabling the voicing of complaints and critiques (Korcynski et al. 2013). This ambiguity is evident throughout this chapter, but captured most acutely in the collaborative playlisting episode, in which a spontaneous communal activity became an obligation. Inadvertently, management revealed the tensions between employers and employees and the latent sense that the latter's much-vaunted 'passion' and 'creativity' were primarily viewed as a useful means to 'manage the feel of work', as Michael Siciliano (2016: 688) helpfully puts it. Organizational theorists refer to such managerial strategies in terms of 'identity regulation': the effort to align employees' personal values and commitments with employers' goals (Alveson and Wilmott 2002). Here, the clumsy instruction to bring staff's investment in 'music culture' in line with 'corporate culture', at first treated playfully, eventually fostered an atmosphere of anxiety, frustration and farce. The self-disciplining effects of the office soundtrack were rendered uncomfortably, almost humiliatingly, obvious – with even an echo of workforce silencing, as when broadcast music, such as the BBC's *Music While You Work* programme, was piped into post-war factories (Korcynski et al. 2013: 203–31). The oft-unspoken rift between these cultural professionals' cultivated identities and the realities of their non-creative task flow was rendered absurdly audible. Unintentionally, albeit on a miniature scale, they implemented something like 'work to rule': a collective resistance tactic in which official regulations are followed to the letter until the normal functioning of work breaks down and productivity collapses (cf. Morgan 1986: 64). The tools of organizational control are turned against the organization itself. The implicit message of such a tactic is: only when we truly become the automata you are treating us as will it become clear that the value we produce for you resides in our own humanity.

6

Standardization

Unbundling the back office

Two buzzwords associated with the 'age of access' inaugurated in the late 1990s and early 2000s were, as discussed in earlier chapters, 'disruption' and 'disintermediation'. The middleman was history; the future lay in fans' unmediated access to the artists they loved. Ownership and remuneration nonetheless remained important and by the mid-2010s, once the digital dust had cleared, it seemed that such innovation had only 'increased, rather than decreased, processes and patterns of intermediation' (Negus 2015: 121): collective management organizations, content aggregators, representative bodies, advisors, analysts and experts of various kinds. One overlooked aspect of this digital re-intermediation is what Melissa Wald (2011: 232) has termed a 'revolution' in 'back-office administrative functions'. These functions are largely understood in terms of the production, management and coordination of rights and recordings, which have indeed been subject to a series of technical and structural transformations. Foremost, digitization altered the chosen format itself: abandoning tangible discs, records and tapes allowed tracks to be isolated, or 'unbundled', from their host 'carriers' – albums, EPs, singles – and enabled them to take on careers of their own, most prominently (and legitimately) through Apple's iTunes. At the listener end, this facilitated a reduced demand for 'bundled' packages, physical or digital, which was not initially offset by the increased demand for individual songs (Elberse 2010).

Ultimately, delivering this at scale required increasing interactions with (or reliance on) technology companies, including digital platforms and consumer electronics manufacturers (Hesmondhalgh and Meier 2018). With such shifts emerged both a renewed set of business models and listening experiences based around the performance of individual tracks as audio content, whether via streaming services or licensed reuse in television, gaming, advertising and branded experiences. Influencing consumer behaviour, financial income, broader ownership and power structures, the implications of unbundling were profound for the industry as a whole. But underpinning the fluidity of informational flows is the stability of digital infrastructure: the need to create, process, maintain and monitor the associated data and metadata, and to do so in line with cross-industry standards and business relationships (Morris 2012; Wald 2011).

For Sonaverse, this implied a need for IT software and processes – and the relevant staff to coordinate them. While the next chapter zooms in on this back-office work, here I focus on how these dynamics played out between the organization and broader industry as new, more complex technical and managerial systems and accounting procedures were being implemented. Beyond the corporate culture of the previous chapter, I also want to shift attention away from the macro-scale on which industry transformation is often discussed – the digitized and disrupted, globalized and financialized, late-capitalist political economy of music – to centre it on organizational bureaucracy and the kinds of everyday activity that are needed to keep it running. I do so with respect to three mundane technologies: first, standardized product codes; second, various kinds of contract; and third, enterprise resource planning systems. While such technologies do shape experiences of music's consumption, what I want to stress in each of these cases is that they also act as crucial *organisational devices*, texturing the conduct of working life inside the major label and connecting it to the demands of broader industry structures outside.

Coding products

Today, the importance of data management to companies operating in a digitized industry, driven by 'unbundled' content and intellectual property regulation, hardly needs stating. Writing at the end of the last century, however, Keith Hill (1999: 1229), then of MCPS-PRS, stands out as something of a visionary when he argues that '*the single factor* which will become recognized as the most influential in promoting the success of electronic commerce' for music and its related media forms would be 'the definition and adoption of standards' [my emphasis]. Hardly a popular topic for everyday conversation, it is difficult to understate the importance of standards. The MP3 format 'allowed for the proliferation of standard objects that could move between countries, media, operating systems, and protocols' (Sterne 2012: 146), for instance, while metadata standards such as the ID3 tag 'brought value, in a corporate sense, to digital music by making it recognizable, sort-able and searchable' (Morris 2012: 856). When it comes to the contested domain of intellectual property rights and royalty payments, product codes and identifiers need to be recognized among these crucial standardizing technologies of digitization. Accounting for music has long been based on the manufacture, transportation and retail of wider consumer goods, where such physical logistics have relied on the Universal Product Code (UPC) – more commonly known as the barcode (Thrift 2005: 221). In the different (but not so different) context of book publishing, Ted Striphas conveys something of the importance of the International Standard Book Number (ISBN) in producing the contemporary field:

> *Hidden in plain sight, product codes have emerged alongside a more familiar cast of human characters (authors, book publishers, agents, booksellers, readers, etc.) to become a vital element in the growth, shaping, and consolidation of the modern book industry. This is not to imply that the book industry would not exist absent these codes; it is to suggest, however, that without them the book industry as we now know it would not exist and that certain actions many people now take for granted (e.g., ordering books online) very well might be impracticable, perhaps even inconceivable.* (Striphas 2005: 273)

In music, this function has long been performed by the catalogue number. These unique codes, assigned to each of a label's separate releases, have just enough visibility to form part of the industry's cultural paraphernalia. Lusted over by record collectors and targeted for ironic *détournement* by punk and post-punk situationists, examples include the ironic commercialism of Stiff Records' BUY prefix for its singles catalogue; or Factory Records' propensity to assign FAC catalogue numbers to stationery, buildings and a sketched design for 'a menstrual egg timer' (FAC8).

At the end of the twentieth century, however, the predominant royalty accounting methods and computer systems, based on the UPC and the catalogue number, were simply not set up to process an informationalized mass of transactions for isolated tracks (Hill 1999: 1231; Wald 2011: 227). It soon became clear that the growth of digital music demanded a new standard. The International Standard Recording Code (ISRC), is a unique, permanent identifier attached to each distinct recording, including multiple versions that might exist of a particular track or video by a particular recording artist, quite separately from its original 'carrier' or rights holder. The ISRC – ISO 3901:1986/2001/2019 to give it its full title – was originally published in 1986,

> *in order to facilitate the accurate exchange of information on the ownership, the use of recordings and to simplify the administration of rights in them. It is a global, unique method of identifying sound and music video recordings [. . . which] enables the tracking and tracing of these recordings through the music value chain.* (IFPI 2009: 6)

Major record labels had already started the lengthy process of assigning ISRCs to tracks in their back catalogue in the 1990s, encountering a number of issues in the process: the linking of multiple ISRCs to a single International Standard Musical Work Code (ISWC) for the underlying composition, for instance; the subdivision and authorship of classical music works; or whether subtly different edited usages of the same track for mixed compilations demanded a unique code each time (IFPI 2009; Elton 2008). When the need to legitimate and monetize unbundled digital consumption accelerated the urgency of this process at the turn of the millennium, the ISRC was thus the closest thing the industry had to a workable, if still imperfect, technology. The IFPI recommended its widespread adoption in 1988, becoming the official registration agency the following year. The standard was revised by ISO in 2001, and again in 2019, with the main benefits seen as enabling the 'interoperation of different databases and systems', readability in a range of hardware consumer electronics and the facilitation of copyright protection in anti-piracy schemes (IFPI 2009: 6). As a result, systems needed to be updated – and, significantly, a growing influx of administrators was required to take care of all the extra accounting.

One such data administrator was Nathan, who joined Sonaverse in the mid-2000s after a stint working at a collection society. Now in charge of the team in control of one major label's product system, he described his remit as 'making sure the repertoire that we own is registered [. . .] based on artists that we've signed, territories that we sign them for, and contractual rights'. The company relies on people like Nathan to oversee the routine entry and maintenance of data (and metadata) across songs, releases and artists, ensuring they accord with cross-industry standards. 'We enter a lot of information', he told me, in order that it can be 'fed externally to the likes of Millward Brown, the Official Charts Company [OCC], MCPS, PPL':[1]

> *We do enter a lot more now than we used to. And that's solely because that information is fed to so many external companies, so it is effectively a library that*

we are constantly updating. Whereas before, we didn't necessarily feed directly to the OCC or anyone else, now we are. So all of that information has to be 110% correct, 110% of the time. So we also need to enter producer credits, or mix credits, as well as the writers and publishers, things like that. Because again, it's all fed directly from this system.

Nathan, interview (2015)

Due to the scale of integration across the value chain, from consumer-facing services to streaming platforms to artist payments, precise data management is critical, albeit hardly glamorous, work. Arranging, linking and maintaining information on writers, performers, rights owners, dates, locations, track lengths, identifiers and so on poses fiendish technical and political problems of recognition, remuneration and strategy for the sector as a whole (IPO 2019). Even less visibly, it also constitutes the daily taskflow of the administrative work that patterns so much of organizational life. And for good reason: the core database that Nathan's team maintains 'drives everything in the business [. . .] if we don't have the product in this system it doesn't go anywhere'.

Contracts and control

The 'back-office revolution' did not drastically disrupt established hierarchies, however. Sonaverse operates a 'federated' structure of decentralized labels and A&R activities, organized around a global distribution and accounting operation. Many employees' day-to-day tasks involve creative decision-making processes: articulating, evaluating, proselytizing, selecting, sequencing, arranging and curating among sounds, styles, images, words and 'talents'. Ultimately, however, everything is accountable to a logic of control, so that much organizational life is made up of more mundane stuff: gaining and authorizing sign-offs, approvals and permissions in accordance with contractual stipulations, managerial hierarchies, procedural checks and production schedules. Negus (1999: 50) characterizes this as a 'loose-tight' management approach, whereby the organizational cultures and routines of satellite labels are relaxed enough to vary hugely within a single corporation, while strategic monitoring and financial planning decisions are centralized and easily 'tight enough to close down or restaff an entire division'. Such a structure was advocated by 1980s business gurus pushing organizational 'excellence', that incentivized entrepreneurial and more customer-oriented corporate 'cultures' (Du Gay 1996: 61). Consequently, at their various managerial levels, 'everybody's targeted, everybody's got their own budgets, everybody knows how much money they've got to spend', as UK General Counsel Tony explained. An international system of 'delegated authority' among executives located in independent departmental 'cost-centres' ensures that on-the-ground decisions – for example, signing a new act, paying advances and royalties – can be relatively autonomous below a certain financial threshold. Corporate governance of this kind is a source of frequent frustration. Reflecting on a career leading up to his recent retirement as a label director, Leonard bemoaned the 'books of rules that senior managers get', restricting his capacity to 'spend a bit more than the remit says, 'cos you're gonna generate more profit', or 'use suppliers that you think are more creative but haven't been approved'. And of course, Tony added, 'as we've become – HAD to become – much more cost-conscious', those thresholds have inevitably lowered over time. In such circumstances, managerial oversight becomes closer, checks and balances more frequent.

All this is to say that bureaucracy and hierarchy run through the music corporation, patterning the micro-politics of office life as much as the power struggles of international relationships. The federated internal structure of the majors is shaped by recorded music's long and complex biography of mergers, acquisitions and divestments (Bakker 2011; Huygens et al. 2001). It remains subject to 'a continual state of restructuring' (Williamson and Cloonan 2013: 16–19): shedding and gaining artists, rights catalogues, business units and staff in cycles. Universal Music Group, for example, reduced its global headcount by over a fifth between 2007 (8114) and 2012 (6422), almost regaining it by acquiring portions of EMI in 2013 (7649), subsequently diminishing more steadily until 2016, whereupon it grew again to almost 10,000 in 2022 (Ingham 2017; 2023a). A bird's-eye view of workforce size gives some illustration of these cycles in the aggregate, but it can also obscure the extent to which telescoping numbers are not necessarily experienced evenly on the ground. All employees are bound to corporate hierarchy, and all sign employment contracts encouraging them to opt out of the maximum hours of the European Working Time Directive and hand their employer a 'right of first refusal' on any artists or innovations they happen to generate – just in case. Nonetheless, stark divisions remain between 'loose' creative enterprise and 'tight' corporate administration, such that employee experiences of this patterning differ enormously, depending on the position within the company. In contrast to those business and creative executives whose roles overflow the boundaries of work and leisure, administrative jobs are predominantly routinized and governed by employment contracts based on a fixed spatial and temporal order. More so than most, admin was sedentary work, bound to a desktop computer in the same office, from Monday to Friday, 9.00 am to 6.00 pm. Interviewed in the wake of the EMI acquisition, Universal's CEO Lucian Grainge cited a need to focus 'job losses [. . .] in the areas of duplication' in order 'to expand the creative parts of the company' ('Lucian Grainge on the Record' 2012). Grainge's depiction of implicitly 'uncreative' parts of the company as 'areas of duplication', and therefore ripe for streamlining, highlights a common view of the back office as a regrettable necessity, to be disposed of wherever possible (Bilton 2015).

Despite such a view, as Tony stressed, the commercial and technical support teams servicing artists have also needed to become much more 'responsive to the needs of the business', in an increasingly complex external environment. In an intensification of trends from the 1990s (cf. Hesmondhalgh 1996; Negus 1999), the relationships of ownership, competition, coordination and mutual synergy that were forged between corporations, artists and other services, through licensing deals, joint ventures or revenue share models, concentrated control into fewer hands. Tony articulated how Sonaverse's legal teams were, by the mid-2010s, constantly negotiating deals 'on both sides of the food chain', as he put it, working alongside the development of both talent and business relationships: 'we've got negotiations on the one hand with artists and producers, but we've then got, on the other side of the coin, negotiations with the people that use that music', primarily digital music platforms. In terms of the former, increasingly intricate contractual arrangements have been important to major labels' resilience, with a growth of contract (or 'deal') types that would once have been considered 'nonstandard' (Guichardaz et al. 2019). This includes the much-discussed '360 deals', where record labels claim a share of a range of associated rights and income streams, beyond those generated by the master recording: live performances, merchandise, sponsorships and so on (Marshall 2013). Equally, it includes more flexible deals based on labels providing an 'a la carte' menu of services, from marketing to distribution, rights management and so on (Homewood 2018). In addition to a diversification of deal types, the sheer quantity of signings also increased: 'There are just so many more deals now', laughed Sophie, shaking her head; 'I was looking at a roster from 2002 compared to a roster now [in 2015], it's like three times as many acts! I mean it's ridiculous!'

On the other side of the 'food chain', the company now constantly develops relationships with new business partners to market, distribute and monetize content. 'Spotify, Apple, YouTube, Google; all of those and many more', described Tony, 'there are hundreds of these things: every day there are new deals coming in for small services'. Commercial Partnerships manager Nick concurred: 'everything is geared to licensing and agreeing deals with partners: number one, the function is to do deals. And, you know, they come up for renewal, there are things that change every time the deal's renewed. So everything is geared towards that'. We can connect this assertion with Nathan's description of the database that 'drives everything in the business': both concern the need to smooth flows – of information and of payments – inside and outside the company. Reducing friction is not easy. Comparing this situation with the relatively static relationships formerly in place with physical retailers, Tony described a process 'a lot more complex than it once was', fraught with uncertainty and speculation:

> *You're trying to work out what the revenue splits between you and the artist should be on business models that don't even exist yet. In the old days, you know, it was a CD and the artist got paid a percentage of the selling price of the CD. It wasn't very difficult, there weren't that many levers to pull. Now when you've got services where, you know, you've got free trial periods, and then you've got different pricing for students, discounting, you've got per-stream minimum rates, and then you've got big advances – you've got lots and lots of different moving parts. And unlike the CD, every digital deal is a bit different. All of the, you know . . . Apple is different to Spotify, is different to YouTube, is different to Pandora. . . . They're all different business models. And you've got to, the contract with the artist has got to accommodate all of those different models, many of which don't exist today but will do. So you're trying to think of all the, every single permutation: 'what if this happens?'*
>
> Tony, interview (2015)

If a popular image of major labels sees them as homogenous, monolithic units strategically differentiating themselves in a competitive market, Tony paints a contrasting picture. From a legal perspective, he draws attention to how Sonaverse needed to balance significant levels of diversity *internally* (a 'federated' organizational structure with a number of distinct work cultures) against increasing levels of symbiosis and standardization *externally* (licensing deals and operational integration with other businesses). The devices for doing so – the contracts of licensing and various kinds of service delivery – had not yet settled into models that could be implemented long-term. The implications of this constant renegotiation of terms and arrangements rippled down into uncertainty and change within the employment contracts, task flows and structures of the workplace itself.

Planning the enterprise

An inheritance of mergers, acquisitions and divestments had made Sonaverse something of an institutional mongrel, causing strategic and technical headaches for the managerial and administrative sides of the business. Negus (1999: 47–50) describes the adoption of portfolio management techniques over the 1990s, such as those of the Boston Consulting Group, as an important recognition of the need to spread investment risks across a range of different artists (or 'products'). In turn, these went on to reshape organizational structures and processes of accountability. Post-millennium, updating the technical infrastructure underpinning firms was a key concern. Most crucially, perhaps, are Enterprise Resource Planning (ERP) systems: organizational software platforms that standardize, centralize, integrate and automate core processes, enabling direct links with suppliers, distributors and customers, as well as close monitoring of product flows and workforce activity. ERP packages such as Oracle, SAP, Microsoft Dynamics or Salesforce have their origins in the problems of warehousing and distributing physical goods and managing supply chains in the 1980s, but today they underpin the core technical and financial operations of any multinational corporation (Pollock and Williams 2009). Offering a range of different 'modules' for different business needs (accounting, stock management, human resources, and so on), they promise packages that are flexibly assembled and customized according to the specific needs of each industry or organization. The widespread adoption of ERP places it at the core of contemporary logistics, to the extent that it is sometimes viewed as the primary driver of a globalized and standardized 'supply chain capitalism' (Rossiter 2016).

Such technocratic visions can seem disconcerting, even dystopian – but they are undeniably attractive to those charged with managing complex transnational organizations. In a webcast for SAP, Dave Cornine, then vice president of Global Financial Systems for Sony Music Group, described how the company responded to dwindling sales by pursuing a rationalization programme of offshoring and downsizing.[2] By the end of the 2000s, they were turning to information systems to further 'reduce costs' and 'gain efficiencies'. The issue, once again, was Sony's federated structure: its multiple accounting systems, or variations of the same system, across multiple national, corporate and market contexts, attuned to the legal and financial regimes of different nations. While SAP was implemented for generic financial processing and reporting, custom modules were required to manage the complex specificities of royalty processing, with different instances of the software used for physical and digital products. Moreover, North American and European operations were using different configurations of the programme – and those offices which routinely operated below a certain revenue threshold elsewhere in the world were not obliged to use SAP at all, instead commonly preferring their own local variants. The upshot was a chaotic plurality of administrative back-offices, technical support teams, system upgrades and reporting procedures, which managers and contracted-in consultants began eyeing up for 'alignment'. This started by agreeing on a common organizational structure – company codes, profit centres, cost centres, chart of accounts and project identifiers – around which a single version of SAP could be custom-built. Since operations in the United States and European territories represented around 75 per cent of global revenues, these were consulted and quickly became the first to implement the new systems regime over an 'aggressive timeline' between 2012 and 2015. In the subsequent roll-out to remaining territories, the consultation phase appears to have been bypassed in favour of instructing offices to 'get on board' by 2018.

Thus 'harmonized', Sony could consolidate its back office and data servers, shutting down local systems and ending staff contracts, with estimated savings of nearly $5 million a year. In addition, head offices sought a tighter level of managerial control and global authority to, as Cornine puts it, discipline 'renegade' accountants' wayward 'egos': no longer should it be possible to say 'this is the way we do it in my country'. Finally, but crucially, Sony's capacity for tracking and analysing data across its various projects no longer relied on generating and combining multiple spreadsheets based on different project codes in different contexts. Instead, SAP enabled a global 'single version of the truth' to be established for its database systems. Cornine explains, using a high-profile example:

> Say One Direction has a new release; they are assigned a unique project code that all countries would follow and use that same project definition. And this enables us now to track all of our costs associated with marketing, promoting and recording and match that up against the sales of that product, so that we can now provide a project P&L [profit and loss statement] at various levels: we can go as high as that release, down to the individual products, down to the individual tracks in some cases, if we want to capture that level of detail.

In the context of a recording industry based more on tracks than album releases, the capacity to 'capture' financial performance in such detail, across markets, territories and company divisions, was significant. The company thus emerged from this exercise more financially secure, with expanded managerial control and greater analytical firepower.

Sonaverse itself can be imagined as a system within an industry that runs on standardization: from bespoke enterprise architectures governing access to departmental budgets or enacting the responsibilities and rewards contained in employment contracts; to product codes and protocols ensuring that the company's communication and distribution flows are harmonized with those of the sector and of international trade. To imagine the corporation as a body (from the Latin: *corpus*), technical accounting systems form a skeletal structure, around which is wrapped a capillary system of managerial relationships. Through these structures flows the ostensibly unifying organizational culture discussed in the previous chapter: an enterprising musical spirit – of expressive individuality and label- and genre-based professional identities – which is encouraged, even mandated, across all employees. As we have seen, assertions of non-conformity and even pockets of resistance remain possible, although typically as a means of negotiating a compromise between top-down instruction and the maintenance of a meaningful work environment for passion-driven employees.

Meanwhile, the formal qualities of popular music – the manner through which recordings are packaged, promoted and consumed; the institutional history of ownership and accounting – demands an integrated set of techniques for administering 'content' (such as master recordings or album artwork) according to specific contractual rights and often eccentric payment obligations and loopholes. Such distinctive idiosyncrasies have fostered a market for home-grown ERP services targeted at smaller media organizations across the industry. The cloud-based catalogue management, discovery and licensing platform Synchtank is one home-grown option; the rights management platform Music Maestro is a more focused configuration of a generic package, owned by Vistex (itself a part-subsidiary of SAP), which also offers dedicated versions for agriculture, chemical and automotive industries. The technological imaginary of these 'off-the-shelf' solutions, however, is their seamless interoperability at a more generic or abstract level, as apparently universal applications that underpin an idealized vision of total automation (Pollock and Williams 2009: 24). In this light, such systems present an oft-forgotten counterweight to the enormous financial and ideological investments that many music actors have placed in disintermediation – of which blockchain and distributed ledger technologies offer a more recent update. While tech enthusiasts continued to imagine a decentralized and automated world, extinguishing the need for trust and uncertainty (Baym et al. 2019), the majors were quietly but firmly reasserting the power of organizational hierarchy.

Systems work

The call

In mid-2014, Kris Weston was irritated. As a former member of the British dance act The Orb, tracks he had worked on in the mid-1990s, while the band were signed to Island Records, were being compiled on a retrospective album by Universal Music Group. Universal had enjoyed a legitimate claim to the tracks since 1989, the year when, as PolyGram, it had relieved Island of its independence. But they had not asked his permission to reuse the master rights in this way. Complaining that he would not have granted consent since it conflicted with a completely different release that he had been planning, Weston made contact with the mega-major to correct this. The exchange was not productive.

> Within 30 seconds of my call to Universal to ask them to immediately cease and desist, some extremely rude bitch told me I had no rights over my work, and that she had the contract in front of her. She sounded self-satisfied and pleased with her day job, presiding over the raping of musicians [sic] work. What a piece of work. She had the contract in front of her within 30 seconds of my call, which I thought strange. When I asked to see the contract it took her a MONTH to give it to me, and yes they properly stitched me up. No control, no legal advice. (Weston 2014)

The full missive, posted to Weston's own website, moved lengthily via creative differences, personal disputes and contractual disagreements, alongside his own struggles with mental health. It presented a detailed and troubling history of a loss of artistic control, amid the raw messiness of human relationships, that is so common to the music economy. Weston's attempts to make sense of that mess etch thick dividing lines: 'creatives' are cleaved apart from their 'non-creative' counterparts; artwork from paperwork; 'lifestyles' from 'day jobs'. The boundaries are familiar – but the manner in which such binaries work to produce sense-making categories, organize people and channel passions can be difficult, anguished, even destructive. In the process, uncomfortable and disorienting relations of power are embodied and voiced in unexpected ways. Among the most alarming of these was the fact that where one might expect a music company, a very different beast is encountered: a call centre. Weston's telling recalls that of many accounts, in which call centres appear as an avatar of alienated work and standardized customer service across deindustrialized and globalized service economies. Against this disembodied bureaucratic dystopia, the voice of the employee who picks up the phone represents not a real person but an abstract system of exploitation, endowed with Kafkaesque responsibility for an array of manipulative mystifications, endless procedures and delayed answers. As is common among administrative positions, that voice happens to be female. Weston adopts the position of affronted customer, and it does not take long for snarling misogyny to surface. Detecting a tone he deems inappropriate, his resort to ugly sexual metaphors associates musicians such as himself with victims of violent abuse, in a somewhat desperate attempt to reverse the imbalance of power.

There can be few who do not recognize this baffling, dislocated experience. Coupled to what we've heard about captive recording contracts and unscrupulous majors, we can assume that most readers of the self-described 'mile long rant' will empathize with the sense of a frustrated artist caught up in the machinations of global corporate capital. While Weston rails against a political-economic system that 'properly stitched [him] up', the real source of his bewilderment and animosity is found in a technical infrastructure that remains differently opaque: the software systems that dominate contemporary corporate life, administrative work in particular. The paper-and-ink of an artist's signed recording agreement must be transformed into the real-world operations that its terms and conditions promise, such as split royalty payments or intellectual property protections. Much of this, increasingly, is automated; much still comes down to human judgement and relationships, or the management of physical space. As Weston discovered – and this chapter explores – the expansion of databases, digital document storage and information retrieval processes in the 2000s rendered contractual metadata accessible at speed; meanwhile, the physical documents themselves are displaced to an off-site repository, so that reading their detail involves the lengthy recall of an archive box and the haranguing of colleagues with appropriate legal expertise. Out of such mundane features emerge those suspicious distortions in time, from the implausible '30 seconds' to the exasperated 'MONTH'.

Weston's 'Orb Rant' circulated on a number of online music news and discussion websites. Reading it provoked me to flash back on the many similar calls that I and my colleagues had often fielded, patched through from the switchboard: calls from artists, their representatives, small independent labels or sometimes members of the public. Often, these were justifiably perplexed as to whom they were speaking with and why, where ultimate responsibility lay or how different departments within the same organization could apparently be entirely unaware of each other's actions. Their own tone could range from the resigned to the abrasive. They could be impenetrable in their own way – as with the veteran artist manager whose reputation for long meandering complaints on anything from the state of the government to faulty domestic plumbing, and just occasionally a disappointing royalty statement, required judicious screening on the part of the administrator. Such experiences are occluded from much popular and academic writing on music. Much attention is paid to how intellectual property is created, disputed, protected and perceived – but not how it is put into action. In popular imagery, for instance, the recording contract appears only in spectacular moments: champagne-popping signing ceremonies and suited courtroom appearances. Between these moments, out of sight, the contract transforms into an 'administrative mechanism' (Du Gay 2004: 39–40) mandating specific functions which constitute a strict division of institutional roles. Hence, the chapter pushes beyond the artist-label focus to emphasize the technologies of organization and the work of systems that go into *enacting* intellectual property at much more mundane levels.

While 'art versus commerce' is commonly understood to structure creative industry tensions, here I focus on those of 'enterprise versus administration' (Law 1994). More provocatively perhaps, I side with the bureaucrats against the entrepreneurs: waged back-office staff, bound by similar employment contracts that encode the expectation of conventional duties and rewards. Power is distributed hierarchically, with workforce dynamics managed through task specialization, promotional pathways and regular performance appraisals (Colbourne 2011). Full-time employees are entitled to standard rights and benefits (including healthcare and employer pension contributions) as well as some 'perks' (free music, tickets or merchandise – although the dwindling of the latter forms a common source of complaint). Likewise, the premise of most internships, or temporary positions assigned by employment agencies, is that they will lead to a permanent position (Frenette 2013). But those embarking on a career at Sonaverse will often be disappointed by the reality of their apparent 'creative' career. Nonetheless, I am not engaged in a moralistic exercise, arguing that admins are worthy of attention because they are somehow more exploited. Sidelined, underpaid and continually subjected to downsizing, they do nonetheless retain positions of relative stability and privilege compared to much more precarious forms of both creative and non-creative work. Rather, I want to highlight the hybrid professional identity that is demanded at the moment when, as employees, passionate music fans come into alignment with technical systems. This hardly makes them the heretofore undiscovered 'true victims'. Mythic tales of exploited artists (and exploitative executives) are legion, not least among record company employees, but quite simply, it remains rare to hear from that voice which Weston describes as 'rude' and 'self-satisfied'.

This chapter listens in to the call from the other end, so to speak, taking us into the back-office world. First, I raise the importance of notions of 'system' in office life: its ubiquity in internal communication, alongside the two core functions of storing and fetching information, and coordinating the projects of distributed individuals. Industry change was, I argue, thoroughly bound up with these functions, which I then explore through the processes of record contract management that lie behind Kris Weston's experience: company archives and their digitization in record-keeping databases. The chapter then turns to administrative labour itself – the kinds of filing and data entry tasks that occupied much of the workforce – arguing that this required both manual work and a form of communicative expertise. Finally, then, I argue that this kind of system work produces a hybrid identity – partly embodying the system, partly relating with colleagues – which underpins a production culture centred on maintaining meaning.

The language of system

Every time I hear a record, I automatically repeat the catalogue number. Music is just a series of numbers to me now. Sales telephonist, EMI, early 1980s (Steward and Garrett 1984: 63).

Walking through the Sonaverse offices, as we did at the start of Chapter Five, we pass displays of gold discs, glossy images of artists and cabinets of merchandise; we are immersed in waves of competing sound sources; sometimes we jostle among established and aspiring acts themselves. As such, we gain a sense of the 'company culture', which is as important for affirming a particular work identity as it is to attract talent (Negus 1999: 63–5). But to *work* in those offices, alongside my colleagues and interviewees, demanded a very different encounter with music. First of all, I needed to learn a standardized language, one that placed me 'among a naturalized landscape of nouns, things, homogeneities', as Matthew Fuller (2005: 97) describes it. Routine commercial decisions and administrative practices represent the stuff of music with a quantified vocabulary of 'assets', 'releases', 'resources', 'repertoire', 'works', 'masters', 'products', 'projects' and 'talent' – terms and objects that signify discrete, repeatable elements. From the global IFPI down to the emailed FYI, the proliferation of acronyms is a sure-fire sign of being enmeshed in a system.

Most of my interlocutors raised the spectre of 'the system' with me unprompted. Understandably so, since software programmes constituted the means through which the operational architecture of the organization – and therefore our working day – was managed. But, more precisely, there were many systems, not all of which spoke happily to one another. The two fundamental processes are *information retrieval* and *project coordination*. In the first, media files, legal and financial documents and metadata structures all need to be stored and organized, so that they can be accessed with relative ease, typically through a bespoke database of some kind. In the second, the various tasks and responsibilities in a project such as a release – legal sign-offs, artwork, marketing campaigns, budget approval and so on – can be managed through workflow software that connects together a division of labour with a cross-organizational, often global, spread. These systems were mainly accessed via screen and mouse. But there is a strange way in which our linguistic abstractions seemed to lend the work more material weight, as something to be lived and occupied.

The very notion of 'system' itself is a 'conceptual technology' (Seaver 2022: 42): an intellectual tool that helps structure relationships between a whole and its constituent parts. It is a way of drawing lines of continuity, both material and imaginative, between technical objects, routinized practices, conceptual frames and structures of feeling which are otherwise fragmentary and dislocated. In the workplace, 'the system' is conjured in different ways: sometimes to name a particular database or technical procedure; at other times connoting, in a looser sense, the general character of corporate life, with its command structures and administrative processes. This slippage speaks to the way in which computational systems intersect with the managerial hierarchies and employment structures which lie 'outside' them, giving recognizable patterns and rhythms to the conduct of organizational life, again enabling some and constraining others. Broader still are the information standards and intellectual property regimes that constitute the infrastructures on which digitized recording businesses rest. In such ways, 'the system' is structured through an array of real and imagined linkages between the industry and the specific software programme.

Drawing from the anthropologist Valerie Olson's study of aerospace scientists, Nick Seaver (2022: 42) refers to this nested quality as 'scalarity', enabling connections to be traced between conceptual structures (such as artists, works, labels and genres; specific contractual terms and royalty repayment rates; or ideologies of freedom and constraint), interpersonal relationships (between juniors and managers, or creatives and non-creatives) and material configurations (meetings, buildings, software and hardware). While informational architectures deal with the ordering of concepts, for admins 'systems' appeared as solid, tangible things – something that new employees would 'have to get to grips with' while veterans would 'come up against'. When colleagues asked one another if they 'have that UPC' or to 'check the PPD on that', it is perhaps not so different from the way in which downloaders once spoke, in a loose way, of 'collecting' MP3s in intangible 'files' and 'folders' – as if to 'concretize' and 'demystify' lines of code into quasi-physical objects (Sterne 2012: 202–3). As such, 'systems talk' is a constant refrain in the major label. 'It's difficult to explain', some might say in response to a query, 'without showing you the system', intimating the extent to which it textures and pressures routine working life. While systems were frequently held to blame for some failure or other, there is a more positive sense in which they constituted the substance of work, lending admins the cognitive and discursive tools to straddle a range of more or less systematized spaces, practices and identities.

Single version of truth

The role of the system is most keenly felt in its absence. Shortly after having left a major label, my interviewee Olive highlighted the lack of systems in her new job managing synchronization rights at a small, and much less well-resourced, independent. Tasked with making decisions over whether or not to pursue a deal – in this case, a song placement in a television advertising campaign – she emphasized how her contributions were predicated on a balance of personal judgements and impersonal frameworks. As a newcomer, this was something that could be difficult and exasperating to manage:

It's not structured at all, there's no database, there's not even a list of the catalogue?
Sometimes I'm like . . . 'How do you know [who owns rights]?' And no one does know.
You have to ask certain people and they usually ask [someone] who retired last year!

Olive, interview (2013)

For Olive, a proprietary 'structure' – an archival store in which internal
company knowledge is sedimented – is a necessary part of completing the task
she is employed to do. In her previous job, answers to such basic questions
were explicitly codified. With no such structure, she is left alone to negotiate
far broader systems of contractual agreements, organizational networks and
global copyright regulation, as well as commercial strategy, by excavating the
tacit, experiential knowledge stored in individuals' heads. Visibly frustrated, she
continued:

Pfff, I dunno how much a TV commercial in Japan costs! How am I meant to
know?! [. . .] I mean, we've got affiliate offices. [. . .] Our US office – obviously the
US is a massive market – but they would be like, 'do you wanna do this?' Like
[exhaling in frustration] – 'do you think it's a good idea that we do this?!'

Olive, interview (2013)

Olive is not advocating the blind following of rules and processes here. The
problem is that such rules are already out there, embedded in broader systems –
of contractual terms, copyright regulations and payment processes, as well as an
unspoken sense of 'how things are done' – to which she has no immediate access.
While her job involves getting 'the best deal' for artists and songwriters, she is
understandably unwilling to act on simple criteria, such as apparent market size,
or with no cost comparison. Divorced of the accessible information structures
of a large corporation, it is hard to determine what makes for a 'good idea'. Each
new decision demands a new epistemological struggle (who knows what?) and a
new valuation (what's it worth?). Nonetheless, the responsibility for the decision
tacitly falls with her to make or influence others' decisions, without her input
necessarily being recognized as such.

[But] I'm the one giving the approval. [. . .] In order to obtain writer permissions
or internal approval, I have to present it and give them my opinion on it: say, 'this
is something we should do.'

Olive, interview (2013)

The final persuasive act of 'approval' is the result of a longer, complex process
of constructing stability and confidence from the ground up. Where her major

label role simplified things down to a handful of button clicks, her job at an indie requires much more extensive efforts in enrolling a range of accomplices – from retirees to colleagues overseas – to assemble historic and local knowledge relevant to each new deal, then secure agreement from other ('creative' or senior) parties before it can be acted on.

Olive's experience demonstrates the organizational value of systems. The knowledge on which decisions rest is externalized in an accessible repository or database rather than locked in an individual's head (the *information retrieval* function), while the work of acting on dispersed authority is outsourced to a system that automatically distributes, tracks and makes individuals accountable for their responsibility at relative speed (the *project coordination* function). Systems produce simple representations of complex realities, in what certain kinds of database design refer to as a 'single version of truth'. By rendering music as a series of 'standard objects', according to agreed technical conventions, uncertain production and organizational relationships can be bolted down, in one way and not another (Fuller 2005: 93). As we have previously seen, the standardization of corporate processes, exemplified in Enterprise Resource Planning, promises a global level of universality, transcending boundaries between distinct 'internal' organizational domains and a global digital economy. Their displays and dashboards involve claims about how the world – or how popular music – 'really works'. Meanwhile, the ISRC standard and its associated metadata enable music recordings to be stored, ordered, transported and used with ease – located in a database, placed in a rights structure, or built into an economic projection – in line with inter-organizational agreements and legal structures. This 'spreadsheet world' (Law 2002: 27) allows system users to occupy an industrial 'reality' far from the passionate attachments and associations through which fans and creatives primarily understand and value music, as well as the disputes over meaning and value that typify much critical discourse.

Keeping records

Systems, then, offer an organizational solution to popular music's vexed relationship with the 'classificatory imagination' (Beer 2013). On the one hand, there is a romantic valorization of music's resistance to order and predictability, finding joy in objects – sounds, tapes, discs and documents alike – half-forgotten, dust-covered, torn, cracked and badly arranged. This latter vision of mess, clutter, disarray, accident and serendipity reminds us that the meaningful stuff of culture is both solid and delicate. In this sense, paperwork is no different to pop. In the mythology of music's material culture, the archive of agreements, statements, memos, recording notes, clippings and cease-and-desist letters in a label's basement is not so far removed from the tattered vinyl records in a second-hand junk shop. What elevates this above mere bookkeeping are those lively oral histories: of contracts signed in a drunken haze and subsequently lost; or decaying company accounts on collapsing stockroom shelves. Talking about Jesus and Mary Chain manager Jeff Barrett, *Creation Records'* Alan McGee (2013: 59) recalls: 'I asked him for the tour accounts afterwards and he dug into his jeans pockets and pulled out this enormous ball of crumpled paper made out of receipts and handed it to me. I'd thought *I* was unprofessional'. Anecdotes of this kind, showcasing a lack of care or a more visceral antipathy to 'professionalism', are one way in which music industry folk and journalists relate a sense of their entrepreneurial passion – via a studied transgression of rules and norms. On the other hand, music is associated with the archival, even bureaucratic, compulsion to systematize: to keep records pristinely filed and arranged in logical order. The chart-tracking, playlist-keeping and metadata-labelling obsessions of online fans and hackers testify to this urge to classify (Hagen 2015; Morris 2012; Witt 2015; Young 2017), finding tentative common ground with the procedural character of modern business – both built on the efficient storage, systematic ordering and rapid extraction of increasing quantities of increasingly complex archives of information. Over a century or more, office workers have repeatedly been equipped with new tools with which to perform these tasks: software directly linking internal processes to global logistics flows is only the latest in a history of bureaucratic innovations, from double-entry bookkeeping to the documents, folders, cabinets and, later, spreadsheets of modern filing (Pollock and Williams 2009; Robertson 2017).

Through the close study and measurement of office labour, as in Frederick Winslow Taylor's scientific management, analysts of industry and employment have long been engaged in formulating more efficient work routines. A central rationalizing principle here is what the mid-century labour process theorist Harry Braverman (1998: 79) called 'the separation of conception from execution': that is, isolating those everyday manual tasks that can be routinized and distancing them from the cognitive tasks of intellectual labour. One rather blunt method for separating conception from execution functions is to distribute them across different spaces. Whether by intention or by accident, such was the case at Rough Trade in the 1980s. Set up and run on non-hierarchical principles, the label and shop were combined, blurring lines between owners and office staff, between making music and packing boxes. In 1984, overcrowding and the need for professional oversight over the label's finances forced a move to new premises, distilling these different kinds of labour into different physical spaces such that they no longer experienced regular contact with, or even visibility of, one another (Hesmondhalgh 1998a: 262; 267).

Something similar was described by Sophie, a legal assistant at an ex-independent subsidiary of a major label. Once acquired, it was common for former indies to remain housed in their own premises in order to retain a distinct organizational identity from the new parent corporation. In the mid-2000s, Sophie recalled, the cost of doing so drove a managerial decision to consolidate real estate and force a 'move from the old buildings, where we had a lot more space', into a single shared building. Inaugurating a period of great instability for the sublabel, this coincided with redundancies and the introduction of new computer systems.

> At that point they were like, 'oh my god, we're moving three or four labels and everyone else into one building, there's not gonna be the room to store all this paper, so we've gotta find a new way of storing everything and everyone getting really easy access to it'. And that's when the concept of [this system] I think was born.
>
> Sophie, interview (2014)

The management of building space is crucial to this rationalization strategy, but what happens to accounts and archives amid such restructuring is not always the foremost consideration. The 'paperless' digital imaginary purports to reduce storage costs and the inefficient doubling of back-office capacities – but, as ever, this was not a smooth transition.

Sophie began by describing the paper-based origins of her workflow:

> Everything has to be logged somehow, it has to be stored somewhere, somehow, in a really easy way for anyone and everyone to be able to access it. So generally, when I first started it was getting my head around that. Just that amount of paper. And the filing systems were . . . Everything was just in a cabinet. And when the artist was dropped you'd send it off to an archive place somewhere that nobody knew about and then hope to god that when you wanted to use it again you could actually locate it. (Sophie)

The stapled paper documents of longform agreements are bulky and bureaucratically troublesome. They occupy extravagant quantities of both space (being alphabetically arranged in filing cabinets) and time (when the fine-print needs to be consulted). Initially, the proposed answer was a simple technical fix: just by scanning the original documents, it would be possible 'to have all the contracts digitally stored somewhere so that [different departments] could get access to it, we could get access to it, we could just send the contracts off to sit in archive'. The aspiration of a paperless office suggests a 'cleaner', more efficient system that would dispense entirely with the manual labour of files and boxes. Unfortunately, 'the scanners they got in place were really, really slow [. . .] there just isn't enough time in the day to do it'. The technical fix turned into a technical problem, requiring a human solution. *Ad hoc* work experience and internships were created 'to help with the admin side of things, the filing and scanning' – but these new members of the team exhibited the expertise, and moreover the care and accuracy, that one might expect of a low or unpaid junior. For such banal reasons, it all 'caused more problems and more work for everyone else in the long run'.

It was clear a more sophisticated (and expensive) system was needed. Digitizing core contractual information as data, rather than as scanned images of paper documents, encoded in a database would be more powerful, enabling the relevant case-specific clauses to be 'unbundled' (so to speak) from the 'standard terms'. Not only would this provide faster access, but it would also enable contracts to be directly linked with other systems based around financial and administrative metadata. Sophie counted seven different systems as central to her process:

> *I've got a couple of systems that I look after [. . .] I've got another colleague that's taken over, there's a new system called [X] . . . So he's taken on those, making sure they're administered. It just means everyone has something that they're responsible for.*

> Sophie, interview (2014)

The gradual expansion and upgrade of such systems have, as Sophie's experience articulates, reorganized the labour process. With this move, the team became, said Sophie, 'totally reliant' on computer systems, to the point that you 'can't really do anything in business affairs without them in place'. Some systems are built from the ground up, others are adaptations of off-the-shelf packages, and users are typically incorporated into the testing and consultation process (Pollock and Williams 2009). 'I question how [management] have decided some of these systems help us as workers', Sophie told me, expressing herself carefully: 'How they've been implemented . . . They've kind of missed the point'. Clearly, it is hardly possible to satisfy everyone's needs, but, in any case, the procurement and design of system upgrades are driven by the top-down imperatives of cost or technical compliance, rather than to facilitate individual talents or work styles.

Programmed labour

Back in the rights management team, requests to license repertoire owned by the company (not unlike those about which Kris Weston complained) occupied a significant portion of the day. Capturing a routine set of tasks, the following extract aims to convey a fuller sense of what is involved in enacting contractual obligations.

> *The marketing team are producing a comprehensive campaign of 'digital bundles', across all sublabels, for the UK. In a folder on my computer are a number of spreadsheets, saved from various product managers' emails, each one replete with hundreds of rows, each row listing track title, artist, label and year of release. These are to be cross-checked against a corresponding data row in an internal database of catalogue (acapella, instrumental, mobile phone download, chorus, clean version, explicit version, album version, remix, radio edit . . .). Collecting the appropriate identifier and ownership data, I can then paste these details into a new product template in a different system – which must then be assigned the necessary release information (corporate division, marketing territories, price point, royalty level, etc). A small yellow button reads 'submit'. I click it and send the project off into the system. Making a note of the project code I can move on to the next release.*
>
> (Research diary 2013)

Rights information is constantly being managed and processed in such ways for many hundreds of projects at any one time. Recording contracts, format conventions and metadata standards have been designed into various systems as 'scripts' – which 'pre-scribe' certain actions and decisions and 'pro-scribe' others (Akrich 1992) – embedded in a host of organizational technologies. Perhaps the most important of these is a generic tool of the contemporary office: a Microsoft Excel spreadsheet. In this case, its rows and columns contain draft track listings, which allows a project manager to play with alternative selections and orderings of the future product based on the pragmatics of availability and cost.

Two further proprietary technologies have been designed around the demands and restraints of managing intellectual property rights for media content. First, a company database indexes tracks as standard objects embedded with unique metadata, providing a global 'single version of truth'. This enables the spreadsheet information to be corrected so that details can be linked into projects. Second, these projects are assembled using a distributed workflow application, linking legal, financial and marketing data, to be consulted by departmental authorities across global and corporate divisions – or (and herein lies Kris Weston's discontent) to be sent to artists and their representatives for reference. As tracks are 'cleared' in this way, the spreadsheet will be continually updated so that, over the life of the project, it becomes the key means of communicating what the new release will eventually 'look like', with relative precision and clarity. It does so in league with other organizational technologies: emails, meetings and phone calls. At various points, the spreadsheet is translated into different standard templates, printed out, passed between hands, and argued over until a final 'version of record' can be produced. As the project is finalized, it is filed along with paper copies of all the communications, decisions and verbal assents in a permanent collection that tracks and evidences due process. Further technologies of storage – cardboard folders, alphabetized in metal cabinets, later to be boxed up, logged in archiving systems and sent to controlled-temperature warehouses – keep the information stable. At least until some future point of dispute or audit arises.

But for now, this is mundane work, carried out alone, according to a technological script which sets the terms on which my hands and brain are set into motion.

The data-entry assignment is relentless but it is one I've been left to complete in my own time. I find myself quickly locating keyboard shortcuts in order to cut down mouse movement. My left hand curls into a claw – thumb beneath palm, fourth finger extended – as I quickly type and switch display windows using alt+tab. The task requires attention to detail so my mind cannot drift – but nor can I focus on a podcast or radio programme. So I listen to music. My headphones envelop me in sound, helping me withdraw from the office and focus on the hands that I'm now watching, as if from a distance, execute cyclical motions, as I abandon the mouse entirely, fluently moving between spreadsheet and application. Involuntarily, and oddly, I recall times spent at the piano as a teenager. For a moment my pace is virtuosic, my fingers flowing in insectile waves. A second of self-awareness breaks the momentum and I slip up, pasting the same information twice. Start again.

(Research diary 2013)

In this compressed depiction of one specific task, two characteristics of major label admin are made particularly apparent. First, this digital economy requires manual labour: embodied informational 'grey collar' work, relying on the repeated motions and energies of people with little autonomy, 'programmed' by particular tools in particular spaces (Qiu et al. 2014: 570–1). Movements are pre-set by established industrial standards and system design, controlled in ways that are familiar from traditional factory and white-collar work (Braverman 1998). It is routine, repetitive and tedious, to the point at which it demands the management of boredom. But second, both despite and because of this, music still figures into the experience of work, engaging and distracting me in equal measure – as histories of music at work show it long has (Korczynski et al. 2013). Indeed, I approach the immersive 'flow'- or 'zone'-like state associated with nominally 'creative' tasks (Banks 2014). For a brief moment, at least.

> *The spreadsheets contain typos and missing details. I search the web to fill in information from unfamiliar artists: it's a mild diversion from the repetition that exposes me to a black hole of hyperlinked information. I didn't know he played on that track. And it was on a Japanese car ad? I wonder what it sounds like . . . More than once does time disappear in this way. The system glitches. I find myself getting viscerally agitated with the software when it freezes for a few moments, making me repeat unnecessary actions or wait on a loading screen. The diversions become frequent and lengthy. I am bored: I find myself listlessly switching between news sites, emails, social media; I fiddle with playlists; I chat with colleagues; I stare; my eyes close. But, after several hours of oscillation between routine and distraction, it feels like I've made a decent dent in the workload.*
>
> (Research diary 2013)

Distractions, interruptions, informal interactions, timewasting and gameplaying are all well-established ways in which manual labourers manage routine drudgery, commonly tolerated by management as techniques for maintaining morale and productivity (Roy 1960; Burawoy 1979). More self-entrepreneurial creative and knowledge workers commonly also find their idealized and aspirational visions of work threatened by mundane realities in office settings. Though unwelcome, embracing boredom can also function to carve out necessary psychological space, protecting oneself from fully surrendering to a work identity that has been moulded by mechanized administration (Costas and Kärreman 2016). Music steps in here too: an involuntary memory, signalling a lingering trace of my former musical self (as a pianist), disrupts a purely administrative persona. Music gives rhythm to my routine, as described in earlier chapters. It acts as a tool to manage the tedium of the tasks while also enabling me to step outside them, furnishing me with a sense of continuity between personal and occupational identities (Leidner 2006).

Hierarchy and hybridity

Notions of employees' skill and value are not fixed but contextually determined, such that recognition and status change over time and space. Increasing levels of functional specialization, such as that experienced by Sophie's legal and business affairs team, shapes work practices and discourses. If this work *feels* 'unskilled', equally, it is institutionally *positioned* as such through the social dimension of the office. Often, this occurs in line with a standard logic of 'deskilling', occasioned by technological advances (Braverman 1998). For digital marketer Poppy, for instance, the customer relationship tools she managed, while intuitive to anyone with basic digital literacy, were vastly at odds with the HTML and analytics skills which had originally gained her the job. She expressed scepticism towards the depth of junior colleagues' understanding, which, in turn, prompted anxiety that her laboriously self-taught craft was no longer valuable. Alongside narratives of technological change, the spatial organization of firms is also crucial. In an earlier office arrangement, in which the legal team was placed at one remove from the label, Sophie had felt that her work was frequently mischaracterized by her frontline colleagues as 'just filing or photocopying [. . .] just putting together documents'.

> We were seen [. . .] as being there to kind of do admin-y, system-y things, and everyone else was there doing their creative stuff. [. . .] I don't think that they've understood the importance of it [in the past]. I think that's the issue. It's only when a photographer is suing you because you've used their artwork and you haven't signed a contract that you realise the importance of your legal team.
>
> Interview, Sophie (2014)

Now seated 'slap-bang in the middle of the A&R department', she enjoyed informally advising colleagues who, sidestepping the system, might 'just walk in and say, "can I just ask you if this is a good idea for a deal . . .?"' Meanwhile, her position in relation to the storerooms and warehouses on which her own work relied was one of relative ignorance: out of sight, out of mind.

What makes the back office of the music industry interesting is the ambiguous and unsettled position between visible and invisible work. Often, this is a departmental matter. Like Sophie, Nathan surveyed the cluster of central functions – 'the IT support teams and some of the digital teams [. . .] royalties, finance [. . .] operations and business affairs' – among whom he was located: 'when you work on a label', he conceded, 'you don't necessarily care about what happens in the so-called back office'. Admins are not handling storage crates, nor do they require the technical expertise needed to manage the 'back-end', only to facilitate constant flows of information. Nathan's team inputs and maintains data and metadata for assets (recordings, videos, etc.), ensuring they are coded and linked appropriately (to personnel, rights owners, dates, etc.). This involves both old and new repertoire:

> We get historic data from our archive, the physical archive. We still have label copy from the seventies, paper copy, that has to be entered in the system because it's never been released in that system before. [. . .] A lot of the data that's in the system is not, as I like to call it, '2015 compliant'! So, if we're re-releasing a product, we have to make sure it's as accurate today as it can be. Because the data that was five, ten years ago, it's completely changed now [. . .] so we're doing a lot of retrospective clean-ups. That's how I spend most of my time actually, doing that.
>
> Nathan, interview (2015)

Although current 'frontline' artists are, he said, the company's 'bread and butter', it is nonetheless 'catalogue' that absorbs most time since it involves a more labour-intensive process of data 'clean-ups'. No member of any other department is able (or obliged) to update the system in this way. Permissions were set only to grant admins access, giving them a certain jurisdiction over 'itchy fingers elsewhere that could manipulate the data to suit their needs' (Nathan). While this may seem to afford admins a degree of power, it reinscribes them in the subordinate position of a relationship between those who *use* the system and those who *maintain* it. A more accessible software interface allowed Nathan's updated database to be experienced as a creative tool by music supervisors, for example, who gained the ability to browse the archive virtually, easily experimenting with combinations of audio and visual media rather than relying on lines of text. Meanwhile, the 'cleaning' and maintenance work that enables such engagement remain obscure.

Until, that is, something goes wrong. The language of system is typically invoked when some object of blame is required, both because it works 'too well – "you can't beat The System" – and not well enough – it always seems to "break down."' (Siskin 2016: 154). Hence systems-based work is commonly understood as a frustrating barrier to action, impeding work – the 'real work', that is, of creative decision-making. Nathan resented 'the perception that we're the police [. . .] here to stop the creative flow':

> *they [creatives] think we are the ones that are stopping the process when we say, 'sorry, we can't do that'. 'Why can't you do that? You can just press a button and change it!' And this is what we're trying to get them to understand. It's not as simple as pressing a button and just magic everything away. It's not as simple as that.*
>
> Nathan, interview (2015)

Nathan's discontent towards a lack of institutional recognition was palpable. 'If it wasn't for the so-called back office, there would be no front office', he argued – incapable of referring to the 'back office' without the qualifier 'so-called' – 'there's no point signing an artist if you cannot deliver a product'. Nathan's tactic in response was to 'build the relationship' with frontline contacts, encouraging his team to communicate clearly and in person, rather than over email. He reported a professional duty to correct misperceptions caused by the use of abstract and impersonal language – essential to 'the actual maintenance side [. . .] but that doesn't mean that we're robots'. Initiates demonstrated their lack of such skill: one, wrongly classifying and filing information by artist instead of rights ownership (as if it were a personal music collection); another, too 'technical', irritating A&R staff with references to arcane system processes.

Communication and translation thus lies at the heart of admins' expertise. There is limited capacity to explicitly influence action: admins relay the progress of particular tasks to colleagues and 'refer up' difficult decisions to superiors. Yet regular correspondence with both colleagues and clients produces a practical familiarity with a range of creative and business disciplines. Consider the following fieldwork reflection:

> *I am in constant conversation. 'This conflicts with our campaign'. 'Push this one please, it's very important locally'. Phones ring, emails are sent, meetings arranged. 'If we clear this for [x], they've said they'll give us [y]'. 'Are you joking, why the fuck would I want to be on this?' Internally, I speak with legal teams, marketers, product managers, label heads, archive assistants; externally, with artists, their managers, their lawyers, other administrators – logging conversations for an unknown future benefit. The weekly project meeting, where calculations are made and decisions taken, offers an education. I have gained a feel for the urgency and viability of various projects, and am beginning to infer from their outset (the money at stake, the artist involved, the executive nurturing a valuable relationship) which will be forced through and which quietly dropped.*
>
> (Research diary 2013)

By developing an understanding of this larger assemblage of decisions and interactions, admins acquire situated expertise. Filed records of previous negotiations and agreements become repertoires for engagement, helping them to recognize competing discourses or value claims and judiciously 'push' (or pioritize) certain projects. Disputes, sometimes belligerent ones, illustrate how good relationships matter, opening space for debate, compromise and partiality across the chain of command. It is not that informal interactions take place *instead* of fastidious protocol, but *in order to maintain it*.

I do not wish to overstate the case: information must be stored precisely; rules must be followed (even as they are open to change). Yet both humans and machines pose their challenges. Computer systems are 'clunky' as Sophie said: they rely on data entry, which may be incomplete or otherwise glitchy. Moreover, while a global workflow application theoretically makes accountability visible, they still rely on executives who are prone to being busy, awkward and unmoved by a backlog of sign-offs piling up in their digital in-tray. Hence, admins might judge it necessary, she continued, to 'go outside the system to get things done'. Similarly, people can be capricious or forgetful. Sophie hinted at the challenges of her position: for example, taking the blame for a superior '[who's] just not in the mood' (as she euphemistically put it, taking care with her words) to give a written approval; 'it's just pressure, you know?' Here, systems provide accountability, durability and security: 'a lot of the artists and their managers and their accountants don't like the system so much', whereas 'from my point of view it's great, it protects me a lot'. To be an admin is not merely to be an appendage of the system architecture but, ultimately, to at least partially identify with it. The hybrid professional identity that results manifests across multiple dimensions: in the capacity to translate between normative 'cultural' discourse and a language of technical procedure or legalese; in an ethical sensitivity to context and why the work matters; or in the thick-skinned resilience to withstand the forthright views and outspoken assumptions of colleagues.

Maintaining meaning

No one is impressed if you type memos or book hotels for a living . . . but if the memo concerns Barry Manilow? Or the hotel room is for Duran Duran? (Steward and Garrett 1984: 63)

The period of change since the turn of the millennium – of digitization and 'disruptive innovation' – is commonly understood at the macro-level of political economy. The reflections and experiences of employees, explored in this chapter and the last, situate the effects of these broad industrial and technological changes firmly in the workplace and establish continuity with earlier epochs of unglamorous administrative and secretarial work. Their accounts invite us to view broad logistical transformations not as abstract problems of top-down supply-chain management but as mundane problems that condition, and are themselves conditioned by, the experience of work itself. Far from the buzzy music culture of earlier chapters, here we began with a scene that could have been from a call centre, moving on to discuss forms of back-office work that are predominantly routine, repetitive, sedentary and mundane. The strict divisions of labour and task function within clear structures of authority and control are all familiar tropes. But there is an important cultural distinction between the call handler, the assembly line worker and those office workers servicing the music business. Put bluntly, the latter *want* to be there because the cultural objects that they are involved in producing, however indirectly, *matters* to them. Indeed, coupled to the promise of secure employment somewhere in music, this may be enough to reconcile many to the dreary tedium of their role. Yet, far from justifying their career in such terms, they tended to view it as a successful *escape* from white-collar work. Thus Sophie, whose law degree taught her that she 'didn't want to be a solicitor', explained: although 'I deal with solicitors, I deal with contracts [. . .] I don't feel it's quite the same as working within a law firm, thankfully'. Admins are conscious that their jobs resemble those in non-creative companies, but the cultural status of the majors muddies such binary oppositions. 'At the end of the day, we could do what we do anywhere', Nathan admitted. The point, however, is that he does not: 'I *like* working here, I *like* what we do'.

To reinforce the point, then, I highlight these workers not primarily to make strong claims about poor working conditions. Insecurity and exploitation do exist but are experienced quite differently from both creative freelancers and musicians and from other forms of routine work in the digital economy, from call centres to distribution warehouse operatives to online content moderators. Rather, their work shows how pervasive systems are in contemporary popular – mainstream, commercial, recorded – music. Systems enable music outputs, encoded in rights and data, to be circulated, monitored, discovered, curated, canonized and valorized at scale and speed across the music industry production system. Office admins are particularly interesting in this regard for their position in the value chain, both close to and distant from the 'action'. As we have seen, the office is a social space with a set of norms and rules and its own vernacular discourse. But it is also a space ordered by archival architectures, organizational hierarchies and logistical networks, in which workers participate alongside desktop computers, scanners, printers, files and filing cabinets, all moving paper and information from place to place. Employees and their practices are connected together according to sets of technical rules encoded in computer software, managerial protocols and standardized forms that traverse organizations and entire industries. These systems shape the doing of work as well as the conditions in which work is done; they shape the language that is used and the professional identities that are formed.

Yet admins are not simple rule-bound conduits for instructions. Mediating interactions between machines and humans, they do not just *transport* information; they *transform* it (Latour 2005: 39), exercising a distinctively situated and hybrid form of interpretive expertise that is difficult to automate (even given the hyperbolic claims made in the name of technological progress).[1] This expertise is not routinely recognized or valued across the organization, however; thus, their work is stigmatized – not just by artists on the other end of the phone but also by colleagues of various stripes. In response, admins defend the value of their work and their position in the system. Equally, however, they must understand the value of the thing they are helping to produce, at a distance, in order to communicate and translate that value in different directions. These system users must possess an orientation towards cultural objects and creative people – described in earlier chapters as a 'passion for music' – which infuses the bureaucratic logic with flesh and feeling (Thompson and Alvesson 2004). In so doing, they hold together the 'motley crew' of heterogeneous, often strong-minded and eccentric, actors in creative projects (Caves 2000). That is, admins actively maintain the dual cultural and infrastructural environment that makes work in this context both possible and meaningful.

Professionalization

Graduate training scheme

Writing on his blog in mid-2008, former A&R executive Ben Wardle was unimpressed by a trip to see his friend Andy, still working in the business. As he walked through the office, something in the air felt different.

> *The faces on the members of staff I passed on my way out today seemed disengaged and kind of, well, disappointed. They were all attractive, young people and had probably punched the air when they landed a job at a legendary place like Warners – many of them had probably got in via the Warners Graduate Trainee Scheme. Yes, Graduate Trainee Scheme: working at a record company is now just another career option, somewhere between accountancy and retail management.*
>
> *'But, what about taking on someone who is already out there doing it, someone with life skills, someone who may not have a degree but has a degree of swagger and originality?'*
>
> *'We'll consider it, but we'd really rather you employed a graduate, Andy.'*
>
> *I can't remember what the outcome was, but it doesn't matter. The proof of who won the argument was in the faces on the graduates sitting in front of computers today. They were probably all thinking: 'Wish I'd gone into the film industry, this is rubbish.' (Wardle 2008a)*

Inspiring Wardle to sit down at the keyboard on this occasion was the apparent disjuncture he saw between the 'attractive, young people' populating a 'legendary' company and the existence of a 'graduate trainee scheme'. In his view, music used to inspire commitment and drive in industry aspirants who were 'already out there doing it' with 'swagger and originality'. Now it has become 'just another career option'. Crucially, the university qualification itself ('possibly Media Studies', he says, but implicitly any such degree) is a warning sign, indicating merely a 'first job on the media career ladder'. Its institutionalization in a formal training scheme is therefore not only incongruous, it reveals the ongoing creep of bureaucratic standardization over creativity – which now extends into individual personality itself, rendered starkly visible in the image of these dispassionate trainees in a dispassionate office setting.

Wardle's blog post captures a real shift in the educational make-up of the industry's workforce in this period, spurred by an expansion of opportunities to gain industry-specific qualifications at degree-level or above. It is also symptomatic of a widespread anxiety that, once music is conceived as a formal profession with established pathways and accreditations, it will lose much of what makes it occupationally distinctive and creatively exciting. Such anxieties found expression in a number of educational, organizational and policy debates, which this chapter seeks to outline. In particular, it assesses widespread scepticism towards the *university* (and academia more generally) as the vehicle for this change: 'Can you really learn how to rock like The Strokes by sitting in a lecture theatre?' as the NME paraphrased (Cloonan 2005: 77). The analogy is glib, of course, not just because it misrepresents universities (reducing them to 'the lecture theatre') but because it misrepresents the industry too (reducing it to its artists). What is at stake here is not *rocking* like The Strokes but rather the work of finding, representing, negotiating, promoting, growing audiences, financing, making logistical arrangements, offering emotional support, collecting payments and sustaining wider infrastructures for not just The Strokes and their peers but their predecessors and successors too. Music is increasingly recognized as a multidimensional ecosystem with long chains of production and complex divisions of labour. It is hardly far-fetched to suggest that maintaining and supporting this ecosystem might benefit from (and contribute to) complex systems of specialist knowledge and expertise. Such normative questions, of what should or should not be the case, must first reckon with the descriptive reality of an influx of music business graduates, which the chapter sets out to explore. The latter is accompanied by a numerical expansion in the university programmes that produce them, alongside a qualitative shift in their content and teaching staff. Indeed, these courses also expand employment opportunities for practitioners – such as Ben Wardle himself, who joined the University of Gloucestershire in 2016 as a lecturer in music business.

Nonetheless, given the complexity of the ecosystem, exhibiting caution is sensible and the question of whether the process of acquiring the necessary knowledge for work in this context should be outsourced to the world of formal academic study is not unreasonable. Perhaps a growth in industry-specific university qualifications indicates a long-term process of *professionalization*. In this scenario, music business management and support work is reimagined as a domain populated by practitioners with relatively high levels of occupational status and autonomy, drawing on specialist knowledge and training to respond to creative, market and organizational problems (Abbott 1988). One alternative to professionalization, however, as Wardle insinuates, is mere *credentialization*: a means of sorting and gatekeeping an overcrowded job market, which in itself adds little value to the performance of tasks or productivity (Collins 2019). Distorted by an expanding mass higher education (HE) system, the handing out of ever more credentials promises to resolve inequalities of access but only reshapes them in new ways (Cottom 2017). Either way, it is clear that a structural shift was taking place here, instituting an increasingly knowledge-driven image of musical capitalism. For the sociologists Daniel Kleinman and Steven Vallas (2001: 451), the growth of knowledge work more widely has been driven by a somewhat paradoxical double-movement – what they call the '"industrialization" of the academy and the 'collegialization' of industrial research' – whereby 'the codes and practices of industry are infiltrating the academy, even as academic norms are increasingly governing the work practices of selected knowledge workers in high technology firms and industries'. It is a neat formulation that helpfully guides the discussion over this and the next chapter, respectively.

While we will go on to explore the development of relatively systematic bodies of knowledge and norms of expertise within and around Sonaverse itself, this chapter first interrogates such a line of thinking through the increasingly close relationship between the university and the broader music industry. It does so by first exploring changes in the British higher education system in which vocational music industry degree programmes became increasingly prominent, as a potential 'industrialization of the academy'. This was a system that I mapped in a study for the industry body UK Music, who were also playing an active part in trying to manage this 'talent pipeline'. I use this context to then explore collegialization, asking what a trained music business 'professional' might actually look like – and what kinds of abstract knowledge systems might underpin their apparent expertise. Turning to practitioners and educators themselves, I note a tendency among both to dismiss 'academic' learning as slow and status-bound (and its 'vocational' marketing as cynical). By contrast, they make appeals to 'the real world' and express a latent will to surface internal knowledge structures: their own curricula, their own training programmes and their own bodies of research – the subject of the final chapter.

Into the triple helix

One way of understanding this shift is to return to the 'creative industries'. This neologism did not simply label a subcategory of firms but signalled a new strategic 'knowledge economy' agenda: an ambition to develop innovation systems driven by, to use a contemporary metaphor, a 'triple helix' entanglement of business, academia and government actors (Etzkowitz and Leydesdorff 1997). The production and transmission of knowledge generated within university institutions were thereby to be enlisted in the employment and investment goals of cultural and creative organizations, underpinned by funding and regulatory direction from the state (Comunian et al. 2014). This was marked by a rise in the perceived need for universities to provide commercial and industrial awareness, professional skills training and work placements across a range of creative and media degrees (Ashton and Noonan 2013). A vocabulary – and increasingly punitive system of metrics-driven governance – arose around the notion of 'employability', a term that suggested a 'good' university course was one that funnelled graduates into jobs within six months. This was a need thought to be driven not just by policymakers or industry practitioners (and their representatives) but by students (and their parents) themselves (Hesmondhalgh 2014: 23–4). A symptom of the broader marketization of higher education, including a more deregulated system, made it easier for new entrants to be established and challenge existing university institutions. With the introduction and subsequent increase of tuition fees, students were increasingly positioned as customers, encouraged to select their education provider from a growing palette of niche options primarily on the basis of 'value for money' and 'return on investment' in the form of projected employment benefits.

If this confluence of actors across the 'triple helix' is difficult to coordinate or intervene in, efforts to do so are commonly located within representative bodies. The establishment of UK Music, with its ambition to represent and support the national industry as a whole, is again paradigmatic here. This trade association for trade associations was built from the work of the British Music Rights lobbying group in 2008 as an attempt to join up thinking and action, funded by a membership made up of representatives of the recording and publishing industries (BPI, MPA), songwriters, musicians, producers (Ivors Academy, FAC, MU, MPG), independent labels and managers (AIM, MMF), and collective management organizations (PRS, PPL). UK Music's remit was to be largely government-facing, lobbying around a number of regulatory and strategic areas: improving the business environment, promotion in domestic and international contexts, training, innovation and so on. For these purposes, they hired a number of senior political advisors with experience in government and set about producing research.

Regular flagship reports documented the industry's contribution to gross value added (figures cited by the UK government as the best available) and workforce diversity, as well as music's impact on tourism and mental health. A core pillar across this broad agenda was support for the so-called 'talent pipeline'. This effluent metaphor encouraged government to consider the lifelong 'skills landscape' that feeds the music industry, from primary, secondary and HE to apprenticeships and recruitment, and ongoing development in venues, rehearsal spaces and formal professional training programmes. In line with their membership, it was geared towards the creative and technical end of the workforce, alongside those in management and support roles. UK Music's 'Skills Academy', a collaborative programme with the government-supported skills council Creative & Cultural Skills (CC Skills), aimed to increase interaction between employers and education providers, especially schools and Further Education colleges, particularly through running careers advice events and coordinating work placements. CC Skills research had identified such provision as wanting, noting a mismatch between a raft of 'apparently highly-qualified individuals' and 'what the music industry actually needs' (CC Skills 2011: 18). And so, by the start of the 2010s, the apparent 'skills gap' – between what education providers supply and employers demand – was firmly on its research and policy agenda.

Within the talent pipeline, qualifications have long been considered useful 'sorting' mechanisms in a market for 'human capital': helpful to employers when 'screening' candidates, who flourish their diplomas and certificates to 'signal' higher-level intellectual capacities and specific skills or interests. But there was a problem here. As cultural economists recognized, artistic occupations are not typical forms of knowledge work and present no simple match between subjects and jobs (Towse 2006). Richard Florida's (2002) conception of a new creative class was also founded, more polemically, on the breakdown of human capital: among what he called 'core' creatives (e.g. neo-bohemian artists, rule-breaking innovators), qualifications were essentially irrelevant. Florida had in mind here the mythical image of the college dropout building a tech empire from a Silicon Valley garage; distinguishing them from established professions (lawyers, accountants) who are tightly bound to certification processes; and a broader swathe of 'skilled' knowledge work (software engineering, financial analysis), more loosely linked to a shifting accreditation system. Such distinctions undercut any notion of a unified 'graduate labour market' that a university might neatly 'serve' (Tholen 2017).

Critics from within the academy contended that a self-defeating instrumentalism was setting in from a government-imposed business model that did not understand how a university produces 'value', within a capped fee regime in which all providers inevitably charged the maximum they were allowed (McGettigan 2013; Collini 2012). In creative arts in particular, the idea of the degree as a period of open experimentation, personal development and exposure to unorthodox thinking was said to be suffocating under high-priced teaching and an excessively vocational agenda (Ashton and Noonan 2013; Hesmondhalgh 2014; Ramsey and White 2015) – especially where the majority of careers were likely to be self-employed and hence neither supported by standard employability programmes nor captured in standard educational outcomes data. Moreover, as I have argued across this book and elsewhere (Bennett 2020a), both those looking for work and those looking for workers seek not just to fill skills gaps but to achieve a cultural match, in the form of tastes, values, identities and dispositions. This is a quality captured by no dataset and understood by few university leaders.

It was into this contested context, at the outset of the formal research programme that underpins this book, that I stepped. I had previously made overtures to the BPI, the main body representing major record companies. Primarily concerned with annual changes in their members' market data, especially the effects of illegal downloading, they politely showed me where to enter my account details to access their library but otherwise declined to engage. By contrast, UK Music's appetite for new research was palpable. They were keen to explore any avenues that would help build an 'evidence base', while my university was actively developing schemes to support pathways towards 'knowledge exchange' and 'impact'. We duly set about devising a project to help them understand the landscape of university-level music business programmes. Their Skills Academy had fostered partnerships with schools and Further Education (FE) colleges; relationships with the HE sector, responsible for undergraduate and postgraduate provision, were somewhat frostier. 'There is a view', I was informed in one early meeting – 'an *extreme* one – that a lot of academic research is . . . ideologically motivated'. In part, this can be attributed to a decade or so of scholarship undermining attempts (notably by the BPI) to lobby for a stronger copyright regime with more forceful restrictions on filesharing (e.g. Cammaerts et al. 2013; Lessig 2004; Oberholzer-Gee and Strumpf 2007). A cautious appetite for change existed, however, with music industry stakeholders having for some years welcomed the prospect of greater collaboration with HE providers and skills councils, considering it a chance to formalize education and training (Dumbreck et al. 2003).

The resultant report, published in 2015, explained changes in the university sector, tracked a rise in specialist degree programmes, and documented some of the caution over this rise among both HE educators and industry managers (Bennett 2015). This coincided with the launch of a formal Music Academic Partnership (MAP), a membership scheme aimed precisely at those specialist courses seeking to broker joint educational and research collaborations, granting students and academics access to industry networks and resources. Universities would be granted MAP membership through payment of an annual fee, enabling courses to be badged with the UK Music logo – an explicit attempt, as they put it, to 'kitemark' courses with an industry stamp of quality. The intention was then to support universities' employability agendas, while delivering an improved evidence base, and so to further communicate music's professionalism – all while helping the political classes to treat the industry as a serious economic proposition.

Growing up

'Professions' are commonly understood as occupations which apply a system of specialist knowledge to specific problems. Implicit in Ben Wardle's account – in opposition to the graduate training scheme – is an imagined ideal of up-and-coming music entrepreneurs in the mould of canonical 'record men': those hustlers with golden ears who report into no one and whose business savvy defies systematization. In a *Guardian* article from the same year, Wardle reproaches Guy Hands, CEO of private equity firm Terra Firma, who depicted the typical A&R at EMI as 'someone who gets up late in the day, listens to lots of music, goes to clubs, spends his time with artists and has a knack of knowing what would sell', from whom power should be wrested away and handed to 'the suits – the guys who have to work out how to sell music'. This was, Wardle (2008b) says, 'designed to hit a raw nerve': a deliberate devaluation of A&R work which would ensure 'EMI will never be taken seriously again'. Indeed, it was an experiment that was widely ridiculed across the industry (cf. Forde 2019; Leyshon 2014). But being 'taken seriously' is indeed what's at stake here – that is, the quest to instil professional legitimacy. In the sociologist Andrew Abbott's (1988: 35) classic account, professions deal with any 'human problems amenable to an expert service'. Many professions require a licence to practice, or at least association with an official body, giving rise to dedicated training and certification mechanisms through formal education: health or legal professions are the archetype here. In Abbot's medicinal vocabulary, they are able to 'diagnose' and 'treat' such problems by means of expert 'inference' – referring to and interpreting complex bodies of academic knowledge.

One effect of a knowledge system is to bestow legitimacy, in the 'mistaken belief that abstract professional knowledge is continuous with practical professional knowledge, and hence that prestigious abstract knowledge implies effective professional work': that is, by referring to 'the research', professionals convey that they are acting in accordance with the higher values of rational scientific progress (Abbott 1988: 54). In practice however, notes Abbott, *problems* are troublesome things that do not simply and uncomplicatedly align with specific experts and their knowledge systems. Hence, there is competition between professions to lay claim to a particular 'jurisdiction', or problem domain: whether alcohol addiction, for example, is a matter for doctors, social workers, economists – or, indeed, music therapists. This is one way to understand Hands' dismissiveness towards A&R: an imperial claim about jurisdiction, making appeals to the superiority of accounting knowledge as a master discipline, delivered by those equipped with MBAs and econometric training. More authoritative, grown-up and serious.

The abstract knowledge systems on which professionals draw, according to Abbot, work by removing practical insights from particular situations, ordering them and classifying them according to a consistent underpinning logic. This allows knowledge to be combined in new forms that prompt innovative interventions (through research) and to be translated into instructional models (such as a course syllabus or textbook). Hence, for him, studying professionalization should involve attending to professions at a systemic level: educational systems that support professional knowledge acquisition; bodies that regulate both education and the application of that knowledge in practice; and the resultant competition among (and within) professions to lay claim to disciplinary primacy. It is to these questions of disciplinary contestation – of ordering and classifying knowledge – within the shifting education system, questions that were of such interest to the British music industry's representative body, that I now turn.

Theory and practice

One of the resources that UK Music housed was the Music Education Directory (MED) – a database of UK FE/HE music courses, established and compiled since 1996 by Allan Dumbreck of the University of West Scotland. These courses were categorized using one of three labels: 'performance and songwriting', 'production and technology' and 'business and cultural theory'. By 2014, according to Dumbreck, the last of these categories had more than tripled in size since 2003, from twenty-six to eighty-four. As with most classifications, this is an imperfect measure for a number of reasons. One important idiosyncrasy here comes with bundling explicitly business-orientated degrees together with 'cultural theory' into a catch-all category of non-practice-based 'academic' learning, more likely to emphasize commercial popular music. But this obscures important distinctions between the 'critical' and the 'vocational' (Cloonan 2005), not least the downward trend of humanities-oriented forms of theoretical training, including in popular music (Warner 2017) over this period, after a 1990s boom – and likewise artificially inflating what would have been an even lower baseline of vocational, business-focused courses. It also undercounts employability initiatives, including additions like work placements or entrepreneurship skills training *within* more 'traditional' or 'technical' music programmes, focused on performance, composition and studio craft – both also on the rise (cf. Cloonan 2005). Thematically adjacent programmes in areas such as creative industries, cultural entrepreneurship or media management (Ashton and Noonan 2013), many of which have music pathways, are also excluded. Nonetheless, the focus on entire degree programmes is loosely indicative – the words 'management', 'business', 'industry' or 'promotion' appeared in roughly half of the programme titles surveyed – pointing to an explosion over the previous decade (albeit one which had started to lose momentum in its final year). Accepting this narrative of expansion, then, the more interesting questions concern the internal contestation between divergent bodies of useful knowledge that it obscures.

'Business and cultural theory' implies two distinct origins – one concerned with investing in and selling music, the other with writing and thinking about music – both straddling divisions of practical and academic expertise, caught up in struggles for legitimacy. 'Cultural theory', it can be presumed, refers broadly to the dominant strand within academic popular music studies (PMS). In the UK, PMS had emerged in the late 1970s from a convergence between music journalism and the tradition of subcultural analysis associated with the Centre for Contemporary Cultural Studies at the University of Birmingham. This was novel and disputed territory within the university setting, involving a fight to be considered distinct from music journalism, traditional musicology and cultural studies (undertaking its own institutionalization) and thereby a field worth supporting and investing in. The goal of one foundational collection of writing on recorded music published at the turn of the 1990s was simply 'to promote popular music theory', with criteria for selecting contributions based on 'a determination to take pop music seriously', showcasing diverse 'ways of studying rock and pop' and 'shifts in method': this was not 'pop history or rock criticism . . . we are concerned with *arguments* about music and its meaning, rather than factual or descriptive material, or judgments of personal taste' (Frith and Goodwin 1990: x). The success of PMS is perhaps indicative from the emergence of 'popular music education' as a distinct meta-field in the mid-2010s, inaugurated by publications, a new journal and an international professional association (cf. Smith et al. 2017). But internal tensions remain.

Anticipating the MED to some extent, Cloonan's (2005) genealogy of PMS splits this 'critical' (humanities-oriented) element from its associated 'practical' (in the sense of creative practice) and 'vocational' (career-oriented) modes of study. Both this and the MED's pragmatic but contentious classification implicitly separate 'academic' from 'practical' and 'creative' from 'technical' subjects. Tensions between 'doing' popular music and its critique or theorization are perhaps self-evident and long-standing. But, as Parkinson and Smith (2015: 99–100) note, the practical and vocational strands of PMS also imply distinct institutional trajectories in the UK. Both have origins in the early 1990s – a transformative period in higher education. First, they identify a growth of instrumental tuition (guitar, drumkit, studio technology, and so on) in private colleges, which, alongside new performance qualifications from providers such as RockSchool, led to the establishment of pop-oriented conservatoires. Perhaps the most visible of these have been the BPI-led Brit School in Croydon, South

London, and the Paul McCartney-sponsored Liverpool Institute of Performing Arts, established in 1991 and 1994, respectively, subsequently joined by the likes of the Academy of Contemporary Music, the Institute of Contemporary Music Performance and the Brighton (later British and Irish) Institute of Modern Music (BIMM). They consider the second trajectory to have been launched at the University of Westminster – a new Post-'92 institution (named after the 1992 Further and Higher Education Act that conferred university status on former polytechnic institutes). A specific commercial music undergraduate programme was established in 1993 as the first of its kind (Parkinson and Smith 2015: 100);[1] to which was later added a masters programme in music business management, alluding to the MBA model (Gross and Musgrave 2020: 135). Experiments in formal MBA programmes also emerged elsewhere: the University of Liverpool's MBA Music Industries (Jones 2002) and an MBA Music and Creative Industries at Henley Business School, University of Reading. Dubbed 'the new rock'n'roll' of business education by the *Financial Times* (Moules 2016), both were subsequently closed and folded into, respectively, a more traditional MA and MBA programme.

Such transformations have allowed the three models – critical-theoretical, practical performance and music business – to spread and hybridize, according to Parkinson and Smith (2015), alongside the staff that have taught on them. Today, it is possible to gain all manner of vocationally driven degree-level qualifications in popular music, designed and delivered by experienced practitioners in upstart 'alternative provider' institutions, many validated in other universities elsewhere. At Tileyard Education, headed by Heaven 17's Martyn Ware, MAs such as music business and international music marketing are validated by Nottingham Trent University and University of Wales Trinity St David. These are set not within a university campus but near to the new cluster of music and tech firms (UMG, Sony, PRS, Google) in King's Cross St. Pancras (see Chapter 3). Similarly, Point Blank music school in East London began as, and remains, a recording studio. Initially offering short technical courses in music technology, it expanded to a number of specialist undergraduate degrees with a focus on electronic music, including an 'accelerated' two-year 'Music Industry Management' degree (validated by Middlesex University), alongside software training in partnership with the likes of Apple and Ableton. Those students looking to combine their study with European travel can take some of its summer courses in the clubbing hotspot of Ibiza, or alternatively, the acclaimed business-focused courses and summer programmes on Berklee College of Music's Valencia campus in Spain.

The real world

In any case, the 'music business graduate' is another relatively new feature of the shift in the music industry's spirit. In itself, this qualification was not yet a recognized signal of applicant quality. Having 'only just got recent experience of those people coming into the market' many managers acknowledged, like Andrew, that 'I haven't thought about them as a discrete group' and struggled to evaluate them as such when recruiting. While employers provide guidance to jobseekers on university courses, they also downplay their importance: 'a music industry-focused degree is by no means a pre-requisite for success', since 'the first thing we look for is relevant work experience' (UMG and Music Week, n.d.: 15; 10). From within the university, it is easy to see how what one educator, Lou (teaching music business at postgraduate level in an older university), described as 'the push from politicians in this country [. . .] to get more instrumental' might be viewed as an unwelcome encroachment on academic freedom. Lou described resistance from students in the traditional university setting who 'do want to be enriched and enlarged':

> *I don't get the feeling that they are instrumental. I never get the feeling that they're, um, just laser-focused on nothing but a job, a career, and getting a paycheque. They want bigger things.*

> Lou, interview (2014)

As a result, graduates of this particular master's programme are 'able to think critically, write critically, have a sense of history and background'. This scepticism towards the 'instrumental' agenda was echoed from the industry side. A&R Manager Zoe, reflecting on experiences of recruitment and as a sometime guest speaker at a university, expressed a broader scepticism towards the employability agenda insofar as, 'because universities have to sell themselves much more' in the new HE 'market', they were incentivized to make questionable claims in course marketing materials to 'prepare' students for work. Pondering the effect this had on school-leavers, she wondered 'whether there's the implication that, by doing this course, you'll get a job'. If so, young people were being encouraged to

> *assume that they need to learn something – but they don't know that they already know it! [. . . The university] just makes people think that there's a right way of doing things and a wrong way of doing things. Or even a respected way of doing things. But there isn't.*

> Zoe, interview (2014)

She considered the form of training offered by a university to be implicitly invested in preserving existing knowledge/power structures through a tactical arsenal of overcomplication, appeals to institutional authority, and a rejection of spontaneity. Lecturers, in her experience, were typically 'older' – or just 'old-fashioned'. To Zoe, the implied legitimacy and superiority ('a right way'; 'a respected way') of learnt 'academic' knowledge blinds students to their own instincts and practices: 'they don't know that they already know it'. Rather than education revealing these unknown knowns, 'it's almost being beaten out of them'.

Seeking a mediating position, Graham (teaching at undergraduate level at a newer university) drew on his own background in artist management and publishing to make the business case for the university. 'The industry's stance is vocational', he complained: '[they say] "We just want someone to be able to do this job"', while misunderstanding that 'experience is not a skill!':

The idea that actually you can equivalently achieve this by the purely vocational practice-base, or versus the purely theoretical approach, I think, is an absolute nonsense. [. . .] We have a vastly superior relationship to our students than most universities have because we adopt this vocational element. You know, we have thought leaders – we've had people from Google, from Amazon, from all around the industry – and we try and stimulate debate and thought but they have a very firm set of nuanced theories that they're able to apply and these are the kind of theories that work on an everyday basis. [. . .] So they're not taught law by people who do it on the back of a cigarette packet, they're taught by people who really, really understand the practice of law. Ditto with the business school with marketing, accounting and finance. They give them a module with several elements of the kind of basic theories of economics. They understand supply and demand, contested markets, competition and so on. So they're given that but at the same time we give them a great emphasis on the practical elements.

Graham, interview (2014)

Graham distinguishes between pure 'experience' and 'the purely theoretical approach' – and indicatively, 'cultural theory' is *not* on his list – preferring instead 'the kind of theories that work on an everyday basis', in practice. His students, by contrast, were not just experienced or qualified but flexibly built for long-term professional success: 'mature [. . .] thoughtful, they're great at problem-solving'. This integrated approach, he contends, elevated his students' learning and the relationship they were able to develop with their tutors, above that of similar courses – which he suggests over-represent one approach over the other (with the 'purely theoretical' implicitly singled out).

Among all my interviewees, and endemic in the wider discourse on the topic, is the sense of an almost impermeable boundary between academia and industry, not simply demarcating different life stages but oppositional *ways of life*. This is perhaps crystallized in popular references to 'the real world', rather than (*pace* my UK Music hosts) the unrealistically 'ideological' world of the academy: slow, bureaucratic, with artificial safety nets. This boundary may or may not be overcome, but the common implication is that it *should* be, since it defines what is and is not possible. For Ben, as an artist manager who had spent several years teaching in one of the new providers of practical and business-oriented courses, the purpose of HE was to provide what he termed an 'accelerated experience' that systematically introduces students to a package of resources – opportunities, expertise and contacts – that would lead to sustainable long-term careers. As such, he registered his frustration with the design of the three-year degree, its attendance and the assessment requirements being out of step with students' work opportunities.

> *There was a time when it was so successful, there would be classes that would be fucking half-empty. You know, half the guys were on tour with some band, or they were doing an internship at a label or an agency [. . . we were] massively victims of our own success. [. . .] And their tutors are being rated on their attendance! The two worlds didn't merge very comfortably at all.*
>
> Ben, interview (2014)

He considered this the consequence of a structural dissonance, between the monitoring of academic standards and the realities of employment:

> *Education funding is based on retention, attendance and rollover – from, say higher diploma to degree, whatever it might be – so the colleges, and the managers of the college, whatever college it is, are under massive pressure to hit their targets in those areas, to continue to receive the funding. [. . .] Basically, what the colleges are trying to train kids to do, they can't let them do whilst they're at the college.*
>
> Ben, interview (2014)

While Ben considered the will to be in place on the part of educators and students, the institutional setting is overdetermined by the 'targets' already deeply institutionalized, *outpaced* by the demands of the 'real world'.

Notably, it is common to narrate this split between worlds in gendered terms. In industry contexts, the myth of the record 'man', and a pugilistic business vocabulary of 'hustling' and 'fighting battles', is pervasive. By contrast, the university environment is often associated with a set of signifiers typically coded as feminine: it is particularly seen as a place of 'nurture', as Graham put it, or the 'developmental personality', as Andrew preferred. The makeup of Lou's course, 'overwhelmingly populated by females', led to suspicions of 'a gender difference in the way these people approach careers':

> *Guys usually – not always – are more aggressive and they're more like, 'you know, I'm just gonna get in the business. I'm gonna start at the bottom and I'm gonna be aggressive about it. What's school gonna teach me? I really wanna learn it the way it is in the real world'. [. . .] Women are more likely to say, 'OK, I'm gonna do my studies first and do it the right way and get prepared and then go in the industry'. That's the impression I get from talking to the students [. . .] the guys just jump in to the industry and do it.*
>
> Interview, Lou (2014)

Again, academia is the 'right way', a sanctioned place to 'get prepared', taking a *care-ful* approach contrasted with the 'real world', requiring a far riskier but more immediate method: 'just jump in and do it'. These are temporal discourses, split along gendered lines: a dynamic, resource-based and atomistic time of *action;* against a patient, reflective and relational time of *care* (Davies 2001). While Lou considered this a neutral interpretation (and framed the university in positive terms more generally), these impressions aligned with reservations over the vocational value of education, arguing that most successful executives had gained their knowledge from 'the streets'.

> *I'm concerned that maybe, for the music business jobs – it's not right, you know? I mean, somebody like a Geoff Travis, or Andrew Loog Oldham, or Kit Lambert, these people didn't have that training [. . .] People like David Geffen, you know – they got in at the bottom level and worked their arse off and were ambitious and learned how to do things and made connections.*
>
> Lou, interview (2014)

The examples, once more, are conspicuously male. A normative implication, then, resonating with both Ben's and Zoe's arguments, is that the aggressively masculine 'real world' does not respect the careful and feminized 'right way'. At best, the academic approach offers an alternative backup for those who lack the passion or self-confidence to 'jump in and do it' alone.

And yet, the apparent impermeability of this division was repeatedly crossed. One effect of the upsurge in music industry degree programmes was the opportunities it offered to experienced practitioners, who were increasingly able to leverage their accumulated expertise. Unexpectedly, multiple interviewees had taught in academic contexts, whether as a full-time course leader or an occasional guest lecturer. For the latter, the university was seen to provide a relatively frictionless addition to a portfolio of (paid) speaking opportunities – a fact that was not infrequently articulated openly to me by interlocutors who mentioned they were 'always happy to come and speak to students'. Practitioners are often valued in the classroom for their capacity to draw on experience and anecdote, telling lascivious life stories, dropping names and sharing conspiratorial insider knowledge. This can give life and context to some of the rather staid forms of theoretical knowledge, with an inspirational guru-like quality sometimes valorized. Legal assistant Melissa recalled an unorthodox, or, in her words, 'batshit mental', lecturer who was nonetheless 'passionate about rights': 'I don't know why they employed her', she affirmed, questioning whether this individual truly belonged in the university context, but she was 'the reason why I'm working here'. As this suggests, however, establishing a career in the university was not always an easy fit, where elements of personality and spontaneity tend to conflict with strict rules around academic standards and safeguarding or curriculum development timelines.

At the same time, interviewees often spoke of a tension, not just from their academic colleagues but from industry practitioners. Graham felt keenly that 'a lot of my colleagues probably look down their nose and go "oh why did he do that?", you know, probably think it's all pretentious'. But, for him, the university had offered a chance to satisfy his intellectual urges:

> *I've often been described as a teacher anyway. I have this need to explain. I want my client to understand why things happen. [. . .] And then, really, the advent of the internet and the impact that had on the industry, I wasn't getting the answers for questions I had from the industry. So I thought, 'I'm going to go into academia'.*
>
> Graham, interview (2014)

Such comments suggest a personal investment in industry's future – finding underlying 'answers' and helping develop creative or business 'talent' – which requires the kind of space and time offered by the university, while posing an implicit challenge to an industry fated to short-termism. My interviewees commonly imagined themselves as custodians and communicators of knowledge: *de facto* 'teachers', even in their industry roles. Although he had moved on from his role in education, Ben spoke of a sustained 'passion for youth development'; meanwhile, analyst Martin felt that his own career successes obliged him 'to speak at conferences, to meet with students' – 'I've got some giving back to do'. Altruism aside, in a context riven with anxieties over age and family commitments, the university might also be seen as a longer-term fallback option – a chance to settle down in (what is perhaps naïvely perceived to be) a less demanding or antagonistic work environment – one, moreover, that offered job security alongside professional prestige.

The old school

Like PMS, then, music business is 'doubly new' to higher education (Cloonan and Hulstedt 2013) – the qualification itself and many of the institutions in which it is predominantly taught are novel developments – both exhibiting significant variation. In parallel with this 'industrialization of the academy' was an insurgent counter-tendency to 'collegialise industry' (Kleinman and Vallas 2001): initiatives beyond the university sector that indexed growing practitioner demand for business training and career development advice. Since the 1990s, in tandem with the 'post-rave growth coalition' of urban cultural industry strategies (Banks and O'Connor 2017), a number of industry conferences and support programmes have proliferated: Generator, a Newcastle-based business support agency and the Manchester In the City event offered localized practitioner-driven alternatives to the university, for instance (cf. Jones 2002). Equally, as increasing numbers of aspiring musicians and small businesses have sought strategies for responding to new and uncertain market conditions, a 'flourishing supplemental industry has emerged', from books to blogs to training programmes, capitalizing on the demand for advice among entrepreneurial DIY musicians and producers (Haynes and Marshall 2018: 463).

Music business courses differentiate themselves from this supplemental industry in qualitative terms rather than on price. Students must weigh up a course's academic rigour, practical experience, historic prestige, industry partnerships, location and so on. We can place UK Music's attempt to 'kitemark' its MAP partner courses, shortly after my HE report was delivered, within this context – as an attempt to use the imprimatur of an industry body to set the terms of professionalization, given that no clear 'winner' had yet emerged from this contested field of university provision. But it is far from clear that there is anything like a stable knowledge base – akin to a diagnostic medical database updated according to the latest clinical trials or a framework of legal doctrine and casebook of precedents – that such a course might impart to their graduates.

Nor, as my interviewees made obvious, is it clear that there *should* be. Without this, whether a formal qualification in music business might be 'taken seriously' as a badge of professional quality remains uncertain. The opportunity remained for industry practitioners to develop their own core curriculum, conduct their own research and train their own subjects. This sets the terrain for the next chapter.

Knowledge work

The New Enlightenment

We can chart our future
clearly & wisely
only when we
know the path which has
led to the present

It's late spring in 2015, and I'm staring at the wall in front of me, trying to make sense of these words. The couch in the company foyer is low, forcing me to stretch my legs out over the artfully distressed wooden flooring. From this height, the mural seems to loom down portentously, its characters fashionably set in multiple 'vintage' typefaces: deliberately incompatible styles, weights, sizes, slopes and serifs, stacked on top of and up against each other. The over-designed period typography of the 2010s – unmistakeably right-now-this-minute – reinforces the historical narrative, flattening past eras into the familiar semiotic clichés of heritage and genre. I'm a little early for my next meeting. Scrolling on my phone to search for this apparent quotation, I locate it first among numerous compilations of 'inspirational' lifestyle phrases, then credited to a 1950s speech by the US politician Adlai Stevenson. I note its appearance in several works of business literature: the epigraph to *Workshift: Future-Proof Your Organization for the 21st Century*, for example, which offers 'a collection of methods that organizations use to align employee work styles and manage goals with business needs' (Morwick et al. 2013: 13). It has been absorbed into the rhetoric of global management – but resonates enough, it has evidently been decided, that it works as an inscription above the gate of the record company workplace.

The uncertainties of the future had certainly been on everyone's mind over the preceding decade, even if, by this point in time, they appeared to be increasingly manageable. 2015 is the year in which recorded music income returned to significant growth. According to the IFPI (2016), digital revenues had overtaken physical sales for the first time, with the rise of streaming subscriptions and licenced services in 'emerging markets' having turned industry's fortunes around. Yet all was not plain sailing. That same year, Mark Mulligan, one of the UK music industry's most respected analysts, bookended the period by publishing his own interpretation of it: *Awakening*. There, he described music's new marketplace of ideas as 'a vortex of controversy and agenda', in which 'the loudest and most impactful voices exercise influence without the need for old world constraints such as qualification or proven expertise' (Mulligan 2015b: xiv). Behind simplistic accounts of traditional media reporting, he continued, 'digital music's various stakeholders [are] increasingly using media to build exposure for their viewpoints':

> *Venture capitalists and entrepreneurs give exclusive interviews to explain that record labels are killing innovation with their financial demands and technology conservatism. Record label executives are quoted explaining how certain services that play ball and are good partners will help build a vibrant future while those that do not are killing the business. Artists appear in documentaries explaining how the old way of doing business is gone for good. Music industry trade associations issue press releases of their latest research that shows how piracy is killing their business while academic institutions counter with their own – sometimes technology funded – research that argues the opposite is true, that file sharers actually spend more. Academics produce literature of varying degrees of quality and authority that provide supporting evidence for virtually every point of view imaginable. (Mulligan 2015b: xiv)*

With some justification, Mulligan's authoritative account positions his own expertise above what he sees as a post-truth melée of competing positions.

But, put another way, it is also an account of digital music's 'economisation' – the term coined by economic sociologists Koray Çalişkan and Michel Callon to emphasize the active 'doing' of economies. Market-formation, they argue, is a process in which 'a multiplicity and diversity of actors compete to participate in defining goods and valuing them' – not just entrepreneurs, capitalists and workers but also those who trade primarily in ideas: 'experts, lawyers, economists, think-tanks and other spin doctors' (Çalişkan and Callon 2010: 8). Mulligan's own interventions, he says, are not only 'evidence-based', rooted in 'research method', 'robust statistics' and contact with hard-to-reach elite executives, but also – and crucially – guided by the 'insight and understanding of the market' born of a fifteen-year career in music, beginning as a recording artist, through to a leading advisor to a range of firms, CEOs, technology companies and financial investors (Mulligan 2015b: xiv). In this way, Mulligan positions himself as an active participant in music's economization while also setting himself apart from the partial 'agendas' of those other actors he lists. His professional narrative thus matches the industry's own 'awakening'; wider still, it matches diagnoses of a more 'reflexive modernity' – one in which societies based on tradition and stable institutions were increasingly being transformed by expert systems and the cultivation of self-knowledge (Beck et al. 1994). In such contexts, argues Thomas Osborne (2004: 440–1), the characteristic knowledge professional is not a technician, legislator or intellectual but a *mediator*: a broker of useful ideas, who can 'move things along' by connecting different knowledges together and transforming them in 'the very buzz of innovation and enterprise'. Music's 'awakening' is a veritable New Enlightenment in which formal knowledges, diverse methods and evidence-based values – cultural insight, statistical data, technical expertise, managerial strategy – have been constructed and put to work. Armed with a command of qualitative and quantitative techniques, Mulligan is able to triangulate numerical abstractions with the tacit, embodied knowledge of his practical experience and musical sensibility. Moreover, his immersion in the milieu on which he comments has lent him a polyglottal capacity to mediate, contextualize, move and translate between an array of disciplinary domains and partisan interests.

This chapter describes a number of ways in which this environment of knowledge and intellectual mediation came to bear on Sonaverse and on those that work within and around it. Ideas, theories, concepts, methods, data and forms of expertise circulate as part of a self-reflexive knowledge world whose boundaries do not coincide perfectly with those of office architectures or employment relations. I conceive of the workplace as one part of what Nigel Thrift (2005) has described as the 'cultural circuit of capital': interconnected networks of executive education, consultancy operations, business publications, industry events and conferences, and internal organizational roles and activities. Having considered the rise of academic music business education in the previous chapter, I initially focus here on the literary world of the corporate music executive: first in the form of trade journalism; second, and more substantially, following recommendations from employees and industry thought leaders, constructing and evaluating a canon of business books. The broader world of data-driven consultancy in which these knowledge intermediaries circulate leads me to consider how expertise manifests within the corporation, concluding with an elaboration of the kind of non-academic training that Sonaverse itself performed during this period. As such, I seek to trace not just the frictionless movement of 'liquid managerial elites' (Davis 2017) through this circuit but also how particular knowledges are taken up and contested within the spaces of everyday working lives – cultivating a facility, among ordinary employees, with 'theory in practice'.

Constructing a professional community

Writing in 2017, BMG's Head of Communications and former *Music Week* editor, Sean Redmond (2017) delivered an op-ed for the newly launched trade journal *Music Business Worldwide*, commenting on the resurgence of this long-standing sub-profession of music journalism. Exemplified by *Music Week* in the UK and *Billboard* in the US, the latter has played an important historical role not only in supplying industry practitioners with up-to-date news and market information but, first and foremost, in offering a meaningful focal point that actually defines and convenes the field itself (Anand 2006). Since their establishment of bestseller charts in the 1950s, these long-running weekly publications have not only documented but actively driven broader shifts in the industry, their fate entwined with that of recorded music. But, through the first decade of the new millennium, subscriptions to the print forms of these magazines had, as Redmond noted, diminished with each restructure and wave of redundancy. Meanwhile, their web-based versions were increasingly competing with a host of born-digital services – titles such as *MusicAlly, Complete Music Update, Music Business Worldwide, HypeBot* and *Digital Music News*; blogs, comment pieces or newsletters from critics like Bob Lefsetz or Mark Mulligan. While the flagship titles continued, many others closed, echoing the widely publicized challenges faced by the popular music press and newspaper industries. In this context, Redmond (2017) saw a resurgent trade press, exemplified by the launch of new print titles, such as the one in which his comment appeared, fulfilling the need for 'an open interchange of information [. . .] in an industry which isn't always renowned for straight-dealing'. In this light, journalists are simply vectors of transparency, supporting executives to make good decisions.

This vision of trade media as a means of disseminating *information* serves to emphasize the importance of *useful* knowledge – knowledge that informs –

targeted to specific audiences for the purpose of guiding action. In other words, it prioritizes a consumer-focused model of knowledge, the logics of which, as Karin Knorr-Cetina (2010) notes, are quite distinct from those of science. Rather than the careful, sedimentary accumulation of facts, generating stable truths about the natural world, market information is engaged in the creative destruction of existing knowledge for a world characterized by dynamism and change. This depiction of the business of music might appeal to those who consider themselves to have honed cat-like reactions in conditions of innovative disruption. Equally, however, Redmond's own position at the 'press officer-journalist nexus' (Forde 2006) reflects the importance of trade press not just in responding to these conditions but also in producing (and reproducing) them. Just as the printing press enabled geographically disparate peoples to reflect on themselves as participants in the 'imagined community' of something called a nation, as Benedict Anderson (2006) famously described, through the regular circulation of newspaper stories, trade journalists work to reflect the business of music back at itself, as if it were a cohesive community. As such, outlets like *Music Week* and *Billboard* remained perhaps the primary means through which the plurality of viewpoints and valuations circulated within Sonaverse. Littered across desks and cabinet tops, those individual, departmental and company subscriptions that did survive ensured the trade press maintained a regular readership within the office. Flicking through the pages of such magazines was a simple way to keep up to date, a sanctioned way to kill time with colleagues, and a prompt for internal conversations.

Both informational and community-building logics are strikingly visible in the practice of media monitoring, in which the significant stories of the day have long been curated and compiled by in-house press teams or intermediary services. The music digest *Record of the Day* (ROTD) has positioned itself as the go-to independent source for the UK music industry from its launch in 2003. Arriving in the inboxes of paid subscribers each morning (as described at the outset of Chapter 5), the email has an uncomplicated design, comprised very simply of categorized lists of links and single-line 'standfirst' summaries, with key actors highlighted in bold. To take a more or less typical example, Monday, 24 September 2012, opened with a track from the indie band Villagers – that issue's 'Record of the Day' – alongside their management's contact details. It recommended gigs that evening for folk singers James Yorkston and Lucy Rose and followed with links to newspaper features on artists both current and heritage, well-known and obscure: Martha Wainwright, Mumford and Sons,

Psy, Brian May, Beth Orton, Philip Glass, Joe Goddard and 'Italian cosmic synth pioneer' Franco Falsini. Beneath this artist focus, the vast majority of its headline stories and opinion pieces referenced Universal's then-recent acquisition of EMI and subsequent divestments. Secondary stories concerned the decline of HMV, *Time Out* magazine's move to a free model, Nick Grimshaw's start on the BBC Radio One Breakfast Show, and a year-on-year drop in viewing figures for *X Factor*. There is a rundown of tabloid gossip – the likes of Lady Gaga, Adele, Flo Rida and Geri Halliwell, among other current and former celebrities – and a host of album, book and live reviews. It closes with 'bigger picture' news stories from the tech and new media world: a high-profile court case between Apple and Samsung; a proposed broadband tax; Facebook's user data tracking activities; and US-led discussions of international intellectual property regulation.

The content assembled by *ROTD* consisted almost entirely of links to material posted in the public domain. Yet the editorialization – the selection, ordering, condensing and pacing of content – is proprietary. The digest was among the first to arrive in the inbox, and on the rare occasion it did not, the apologies from editors were profuse. Just as with the competing slow/fast temporalities described in the previous chapter, the entropy of informational business knowledge is explicit here, since its value lies in its currency and timeliness. Hence, it was common practice to start the morning with it and its competitor services: each day brings fresh stories; yesterday's content is old news. Equally, however, it acted as a way of easing the transition into 'work' proper. Reading its stories helped mediate between everyday workplace routines (emails, meetings, data processing tasks) and broader market rhythms (midweek charts, quarterly sales figures, annual awards ceremonies). Given this, we should ask ourselves, as Anderson (2006: 33) does regarding national newspapers under 'print capitalism': 'Why are these events so juxtaposed?' Since 'most of them happen independently, without the actors being aware of each other' – Samsung may be unmoved by the music of Franco Falsini, who has never had cause to listen to the Radio One Breakfast Show – 'the linkage between them is imagined'. Imagined but not illusory: assembled in a routine email sent to a delimited readership, this collection 'makes sense' – quite literally, it *creates comprehensibility* – and it is this imagined order, upon which a community might be assembled, which is 'owned' and sold by the *ROTD* team. As a note at the foot of each edition makes clear, 'Please respect the amount of work that's put into this newsletter by not regularly pirating it'.

This is not to say, of course, that media of this kind automatically reflect industry consensus. An academic's outsider reliance on business reporting might reasonably be questioned if they were to naïvely treat it as a faithful mirror of industry practice and belief (cf. Wilkinson and Merle 2013). Similarly, such narratives are also challenged in the workplace. As one respondent mockingly countered, in regard to my simplified account of long-term industrial change, for example, 'Well, if that's what it says in *Music Week!*' Reframing my question as a somewhat hackneyed narrative, my apparent remove from the workplace was deftly – and somewhat embarrassingly – exposed. Yet it was also an invitation to a conversation, opening up a space for debate. Such rhetorical moves are common, as insiders joust with their command of industry knowledge. Accordingly, in the office, trade media prompted the sending of emails to colleagues, picking up on a particularly interesting story or ridiculous piece of gossip. Opinions and emerging stories re-circulated through the 'ambient journalism' of social media channels (Hermida 2010), especially Twitter, where insiders with large follower counts comment on current events; in response, workaday employees post their own replies. In the pub, a heated conversation may start: 'Can you believe what [X] said about that campaign?' In this sense, trade reporting is perhaps best considered a conduit for reinforcing a sense of what is important and relevant, enolling its audience as participants in a professional community engaged in a serious project (Edwards and Pieczka 2013). It provides its readership with discursive resources: narratives and interpretations that help make sense of the professional world and frame its internal disputes. 'Market information' bleeds into common sense.

Deep texts

Business media and the trade press in particular form part of recorded music's 'deep industrial practices', in John Caldwell's (2008: 346) terms, fuelling reflexive sense-making within the semi-public world of the 'production culture'. For Caldwell, the circulation and interpretation of knowledge in the form of written texts helps circumscribe what it means to be embedded in that production culture, as an 'insider'. He categorizes this knowledge in line with three levels of 'depth', or intended audience. At the highest level of visibility, 'publicly disclosed' texts are those which the industry produces about itself for popular consumption: websites, making-of videos, fan conventions, PR communications and so on.

These propose to pull back the curtain for those interested enough, giving them a sense of the hidden work that goes on behind the scenes. Trade journalism lies a little further down the scale, alongside electronic press kits, industry conferences and job ads. These forms of knowledge are 'semi-embedded': available to the public but, given their specialist nature as communication among professionals, unlikely to be of general interest or relevance. Finally, he considers some texts, such as software packages, newsletters, meeting agendas, promos or certain forms of gossip, to contain 'fully embedded' knowledge, comprehensible only to workers, as part of routine professional practice. This depth continuum is useful for considering the role industrial reflexivity plays in constructing a community of practice with a particular set of relationships to the audiences it serves: inviting some in and giving them privileged access to insider knowledge, while others are kept at a distance. Nonetheless, it would be misleading to think of this knowledge as a thing that resides *inside* a written text – as if industrial reflexivity and professional expertise are produced simply by acquiring the 'correct' (sufficiently deeply embedded) piece of information. It is easy to make such a mistake. Indeed, this would be to follow the orthodoxy, consistent with the normative coding of 'intellectual property': ideas passed around in container boxes. Instead, Caldwell's taxonomy emphasizes knowledge as a *relational practice*. That is, reflexivity is a form of work in itself, not just a way of communicating across the 'industrial borderlands' but of carefully erecting, maintaining and sometimes playing with those borders (Caldwell 2003). Although implicit in his account, we can push this point further, placing analytic emphasis not just on what texts *contain* but on what is *done* with them.

To illustrate this, I turn to a critical practice and a textual form that he, perhaps surprisingly, barely considers: book recommendations. This is because if I *really* wanted to understand how Sonaverse works, alongside the wider industry it's part of, as I was often told by my colleagues and interlocutors, then I needed to *read books*. This injunction came via more or less explicit references in routine conversations, along the lines of: 'you should read [X title]'; or 'It's like what [Author Y] says – do you know that book?' After a while, I began noting down such recommendations. Then, in 2015, my regular consumption of business media directed me to two further booklists, curated by industry experts and commentators. First, Eamonn Forde (2015), a trade journalist, published a list of 'The 15 Music Industry Books You Must Read' on the industry website *Music Ally* (I name this List A: 'Journalist'). Second, Will Page (n.d.), a professional music economist, published a personalized list using Amazon's now-defunct

Listmania function, titled 'Rockonomics' (List B: 'Economist'). Two widely respected professional intermediaries, their line of work involves explaining complex topics (economic theory; internal business eccentricities) to non-specialist audiences. They are, as such, influential 'thought leaders', potentially even 'public intellectuals', within their field. Both have authored their own books and are regular faces at industry conventions, panels and seminars discussing music business issues with an insider audience. It is worth, in other words, taking seriously their role as curators of ideas – and the process of assembling, translating and interpreting is far more interesting here, than the ideas per se. By this time, my ethnography had generated nine personal recommendations (List C: 'Practitioners'). This process assembled a corpus of forty-one recommended books. The two professional curators added their own brief commentary to their texts, justifying their choices and giving cues on how to read them. Practitioners, on the other hand, tended not to package their suggestions with such rhetorical gestures; generally, they preferred books to speak for themselves.

In effect, this process follows in the footsteps of Boltanski and Chiapello (2005), who mined similar lists of popular business literature to reveal the underlying 'spirit of capitalism'. Their claim is based on the assumption that such books supply their readers with an aggregated conceptual armoury of ideological resources and justifications that they take forward into their managerial practice. Some of their methodological choices are suspicious, however: the mere appearance of a book in, say, *Harvard Business Review* is enough for Boltanski and Chiapello to accept its influence. But this does not necessarily translate into it actually being picked up and read, let alone acted on, by practising managers, nor does it account for the contextual specificities of particular industries and nations (Parker 2013: 128). The list I constructed is more specific, located in the UK's music business of the mid-2010s, and uses recommendations *from the workplace itself* as something of a check on the predilections of a particular business elite. This produces an idiosyncratic literary canon that is found in no library, professional handbook or university reading list. Instead, it is a document of how particular ideas are valued within the field at a certain time and – given that a list is a way of packaging knowledge so that it travels easily (Young 2017) – how they circulate.

This exercise allows us to explore the kinds of knowledge that circulate a little closer to practice. To delineate the broader reading list, after reading and familiarizing myself with every item, I classified each according to genre and subgenre, loosely in line with the intended audience interest. The first genre, 'Textbooks', describes those intended primarily for insiders, presenting explanatory field manuals (I mark these as 'How-to Guides'), technical explainers ('How it Works') and scholarly analyses ('Academic'). These specialist texts – such as the economics textbook, *Statistical Forecasting* (Goodell Brown 1959), or *The Pyramid Principle* (Minto 2008), on business communication techniques – are unlikely to be of interest to a general audience. On the other hand, enthusiasts *are* likely to read record label histories and executive biographies, and even passing consumers may have a strong interest in the financial aftermath of the Beatles' split, depicted in *You Never Give Me Your Money* (Doggett 2009), or *The Song Machine* (Seabrook 2015), detailing the writing process for contemporary hit songs. These are therefore captured in a second genre, 'Popular', containing non-fiction works intended for an interested lay reader; it comprises 'History', 'Journalism' and 'Biography' subgenres. Works of 'Fiction', a third (more self-explanatory) category, are typically marketed to a mass audience and hence the most public. Like Caldwell, therefore, I classify the texts mindful of the professional or public audiences that they address. To these, I added an indicative theme (music, e-commerce, general business, etc.) and noted their edition and year of publication.

For illustrative purposes, the table below shows the seven texts which appeared in more than one corpus subset, plus two examples of the 'How it works' subgenre (from the Economist and Practitioner lists), one of which also gives an example of a subject field imported from outside music. Across all three lists, these books act as signals, positioning the recommenders as knowledgeable authorities, experts enough not only to say 'read this and not this' but 'read it *in this way*'. This is a crucial point about recommendation. These books are not always used in the way that their authors or publishers intended. *All* of them, even novels, are recommended for their pedagogic value: as sources of learning. A practitioner may read *You Never Give Me Your Money*, as Forde writes, not for the 'history and cultural impact' of the band (since there is a veritable sub-industry of Beatles literature) but as an attempt 'to untie the complex business and legal knots they left'; while *The Song Machine* is 'a prescient reading' of 'mainstream' consumption habits. My subsequent discussion explores how the practice of recommendation serves to blur genres and remake readerships, constructing a new canon of trusted knowledge. The degree to which this knowledge is, as Caldwell puts it, *embedded* in the production of culture, therefore, is determined less by the texts' content (the ideas they 'contain') than how a particular value community positions itself in relation to them: as established practitioners, aspiring entrants or consumers happy to maintain a distance.

Table 2 Extract from corpus of business literature recommendations. See Appendix 2 for full list

Corpus Set	Author	Title	Year	Genre	Subgenre	Theme
A, B, C	Dannen, Frederick	*Hit Men: Power Brokers and Fast Money Inside the Music Business*	1990	Popular	History	Music Business
B, C	Harrison, Ann	*Music: The Business. The Essential Guide to the Law and the Deals (6th ed.)*	2014	Textbook	How-to Guide	Music Business
A, C	King, Richard	*How Soon Is Now?: The Madmen and Mavericks Who Made Independent Music 1975-2005*	2012	Popular	History	Music Business
A, B	Knopper, Steve	*Appetite for Self-Destruction: The Spectacular Crash of the Record Industry in the Digital Age*	2009	Popular	History	Music Business
C	Kusek, David and Gerd Leonhard	*The Future of Music: Manifesto for the Digital Music Revolution*	2005	Textbook	How it works	Music Business
B, C	Niven, John	*Kill Your Friends*	2009	Fiction	Novel	Music Business
B, C	Passman, Donald	*All You Need to Know About the Music Business (8th ed.)*	2014	Textbook	How-to Guide	Music Business
B	Szymanski, Stefan	*Playbooks and Checkbooks: An Introduction to the Economics of Modern Sports*	2009	Textbook	How it works	Sports
A, B	Yetnikoff, Walter	*Howling at the Moon: The True Story of the Mad Genius of the Music World*	2004	Popular	Biography	Music Executive

A new curriculum

Only one book appears in all three lists. Fred Dannen's (1990) *Hit Men* is a popular history of the US record business in the excessive heyday of the 1970s and 1980s. Forde praises its line in 'investigative journalism' that treats its subject matter at 'a much more intellectual level' than mere 'salacious tabloidization'.

Page is more concise, elevating business connivance to an entry-level university undergraduate course: 'Music Industry Corruption 101'. Its title, consciously conflating record executives with hired assassins, places it in a similar dramatic space to John Niven's novel *Kill Your Friends* (Niven 2009), discussed in Chapter 4, which Page also considers to 'teach you a lot about A&R'. The latter is one of two works of fiction on his list, both of which are chosen for their instructional value and for their origins in real events. While Niven's novel is informed by his A&R career, the film *Barbarians at the Gate* (1993), a dramatization of a moment in financial trading history, is based on an acclaimed work of (non-fiction) investigative journalism. If the movie has little to do with music and 'feels more like a theatre play', Page enthuses that it will nonetheless help the reader 'understand financial economics, the role of debt as an asset and the relevance of egos', as well as 'the madness of bidding wars' – all of which constitutes a 'perfect primer for the music industry!' Just as the line between fiction and non-fiction is opaque here, our curators also appear uninterested in distinctions between scholarly and vernacular knowledge. *Black Vinyl, White Powder* (Napier-Bell 2001), for instance, ostensibly a factual historical account, is narrated through a series of anecdotes from Napier-Bell's personal biography as manager/record man-cum-documentarist, blending tall tales with footnoted research. In a sense, this is familiar terrain for the 'hybrid style' of business literature, where 'the author is projected primarily as an expert guide (and all-knowing "guru") but with marginal traces of the academic researcher' (Chiapello and Fairclough 2002: 200).

The ubiquitous *Hit Men* is also distinguished as one of the older entries in a corpus where, overall, publication dates skew towards the present day: seventeen of the forty-one were published between 2010 and 2015 and sixteen in the preceding decade. Perhaps there are simply more books these days – publishers may be meeting a growing demand for such literature – but the sample is hardly representative of changing market trends. More interestingly, this recency bias tells us something about what kinds of knowledge our curators think *endures*: those chosen few books that remain relevant enough to sit alongside the 'fast' knowledge of recent publications. Of the eight selections published prior to the millennium, for example, four directly concerned music. One of these, *Psychotic Reactions,* is a compilation of pieces by iconic 'gonzo' music journalist Lester Bangs (1987); two (*Hit Men* and *Rockonomics* (Eliot 1996)) are popular histories of the music *business*; and all three of these broadly centre on themes of excess and exploitation. The fourth music book (*Producing Pop* (Negus 1992)) is a more sober sociological account of the record business, referred to by Forde as '(rightly so) required reading on all music business courses'. Negus' is one of three 'Academic' texts among the eight older texts, alongside *Statistical Forecasting* and McCluhan's (1964) classic media studies manifesto, *Understanding Media*. The latter is an interesting inclusion, recommended by Page for its distanced historical perspective. Having been 'written twenty years before the PC revolution and thirty years before the rise of the internet', it remains useful for 'rethinking where the future has taken us' [*sic*]. McLuhan is retroactively positioned as a business guru, lending his book explanatory 'How it Works' value, alongside *The Innovator's Dilemma* (Christensen 1997) – the classic (if academically discredited) business tome defining disruptive innovation (Lepore 2014) – described as a book that 'has stood the test of time (and continues to get cited)'.

But it is not just these few that have passed that test. Several of the selections have been issued in multiple editions. By 2015, two books, *All You Need to Know* (Passman 2014) and *Music: The Business* (Harrison 2014), were in their eighth and sixth editions, respectively. These are frequently referred to as perennial classics (e.g. Graham: 'take Don Passman's *All You Need to Know about the Music Business* – bin that, actually, because it's about the American industry! [...] Instead you go to Ann Harrison...'). The multiple editions of these How-to Guides attest not just to their established authority (and hence the seeming validity in their insights) but also to their ability to instruct the reader by 'refreshing' current knowledge. The greater number of How-to and How it Works guides appear post-2000, dealing with case studies and situated knowledge, which, it might reasonably be presumed, are more quickly outdated. By a different measure, the past is also valued in terms of the stories told about it – and the more recent texts tend to favour historical subject matter. Nineteen texts in the corpus are historical or biographical in nature, with the vast majority (seventeen) published after the millennium. Five of these concern companies (Creation Records; Warner Music; Napster; EMI; MTV) and two executives (Andrew Loog Oldham; Walter Yetnikoff). Ten, however, are broad overviews of 'the music business', either in its entirety or a particular segment (the 'record industry'; the 'concert industry'; the 'hip-hop business'; 'independent music'). Forde enthuses about the need to take 'the long view' in order 'to show the music industry has fought the same battles and made the same mistakes for far longer than you presume'; they are valued as a 'modern morality tale'. Authors are lauded for their ability to 'breathe fresh air into a story' and to unearth 'solid, polished pearl[s] of new information'.

A corpus like this, then, tells us more about longevity than publication trends. Two forms of knowledge seem to 'stand the test of time', settling into what appear to be universal constants or objective truths about the industry: those books which depict the 'primal' irrationalities of this business (*Psychotic Reactions*; *Hit Men*; *Barbarians*; *Rockonomics*); and those which reveal underlying 'theories' or 'rules' in a more scholarly register (*Statistical Forecasting*; *Understanding Media*; *Producing Pop*; *The Innovator's Dilemma*). The titles alone are indicative here. Echoing the words of Adlai Stevenson, the past is viewed as a learning resource to be mined, retold and reinterpreted from the vantage point of the present, in order to 'chart our future clearly and wisely' in light of 'what we know now'. Far from simply a list of the most recent management fads and fashions, then, our recommenders are concerned with constructing a historical canon that places their expertise within a sense of industry heritage. Such a positioning strategy is not a neutral listing exercise. It is one that helps to articulate their particular brand of intellectual capital.

Finally, then, by focussing on distinctions between the three lists, this classification scheme sheds some light on differences between performances of knowledge within different professional milieux. For example, fifteen of the forty-one books are 'Textbooks' – but only one of them, a work of academic popular music studies (Negus 1992), appears in the 'Journalist' list. The majority come courtesy of the 'Economist' list, overwhelmingly populated with generic business 'How-to' or 'How it Works' guides. Their hyperbolic promissory titles often make claims about 'the future': to 'change the way you do business', to offer secret 'hidden' or 'insider' knowledge, to explain 'how x does y and z'. Described by Page as 'a bible for many, unheard of to many others', for example, *The Pyramid Principle* is explicitly ascribed 'sacred' insider status. The audience is implied in his gloss of one book as 'polishable in a long haul flight'. The popular 'airport paperback' version of management knowledge, claims Peter Fleming (2015: 76), is read by 'very few "real" managers'; rather it is written by executive guru figures for others who aspire to be like them.

Equally significant are a number of books that popularize the field of behavioural science (Heller 2008; Poundstone 2011; Schwartz 2005) – wherein insights from economics, law and evolutionary psychology intertwine to explain human social interaction (or consumer decision-making) without appealing to the fuzzy notion of 'society' (cf. Milonakis and Fine 2009). Tellingly, removing the Economist list causes this subset of behavioural business knowledge to disappear entirely. The remainder of the How it Works genre is much more specific in its emphasis on music and its dedicated industry (counting Menn's (2003) story of Napster). By contrast, while both the Journalist and Practitioner lists include biographies of executives (*Stoned* (Loog Oldham 2001) and *Howling at the Moon* (Yetnikoff 2004)), they are dominated by Popular and Fiction texts – more palatable to the casual reader, perhaps. The lists also reveal their different temporal emphases. Six of the eight oldest publications are on the Economist list, while its more instructive tenor appeals more to universalized rules-based knowledge. Conversely, the majority of the books dealing with historical *subject matter* come, unsurprisingly, from the Journalist selection, indicating a humanistic need to keep the past open to new interpretations. Interestingly, this selection overlapped far more with the practitioners among whom I circulated.

Given the numbers involved, such distinctions can only be indicative. But the choices in these lists do signal the distinct 'cultural circuits' (Thrift 2005) of the music industry's internal knowledge economy, through which 'thought leaders' – their particular brand of influence and their potential audiences – move. Importantly, they reflect the subtly distinct disciplinary traditions and career structures of their curators. A self-described 'freelance music business and technology journalist', for example, Eamonn Forde's trajectory articulates the close relationship between popular music studies and music journalism in the UK. He holds a PhD from the University of Westminster (a study of the UK music press (Forde 2000)), working with PMS scholars Dave Laing and Keith Negus (Forde 2000: iii) – the latter providing the one academic text of his fifteen. Forde moved away from academia to edit the market research publication Five Eight for the branding agency Frukt between 2002 and 2008, delivering regular overviews and in-depth strategic reports for a readership of industry executives. Now a respected trade journalist, he regularly writes for both specialist and generalist publications – such as *Music Ally*, the web-based venue where his list of recommendations was published.

By contrast, sitting in the borderlands between music, data and finance, Will Page is an economist educated to postgraduate level who has worked as chief economist at PRS and Spotify. He has also moonlighted as a music journalist for *Straight No Chaser*, a magazine specializing in jazz, soul and Black music, but built his reputation primarily on a series of widely-cited reports written while working for PRS, 'adding up' the industry and using music to explain tenets of economic theory. His use of Amazon's now-defunct curatorial function, *Listmania*, suggests he is hardly unaware of this public explainer-influencer position. His subsequent appointment at Spotify is likewise emblematic of the streaming platform's efforts to identify itself as a key agent in the music economy's reinvention, with a speculative business model predicated on growing its financial market valuation (Vonderau 2019). Page's professional currency therefore reflects a wider 'empirical turn' in economics trainees: the growing prestige of applied economics careers in business contexts, in which economic historians Roger Backhouse and Béatrice Cherrier (2017: 2) tell us, 'the frontiers between the academic world and IT firms such as Google and Microsoft are increasingly porous'. Tellingly, Google's chief economist, Hal Varian – an exemplar of the empirical turn – provides Page with one of his key readings.

Equally, the list of organizations they are associated with tells us something about the evolving circuits of, in Thrift's (2005) terminology, a more 'reflexive', 'knowing' musical capitalism. *Music Ally* positions itself as 'a knowledge company' at the intersection of 'music' and 'tech'. It offers a portfolio of services, including professional training, events organization, marketing, news, and consulting, drawing on 'over 15 years worth of information, experience and data on the global music business' to advise clients 'from government agencies to startups and from music platforms to labels'.[1] PRS for Music, a much older institution in the industry at the national level, works to pay out royalties related to a range of recording and publishing rights. Alongside a regulatory and administrative role tracking the exploitation of rights and royalty payments, PRS in particular takes an interventionist approach to 'supporting creators' (and hence market innovation) through financial support, business advice and lobbying (Street et al., 2018). Sitting at the confluence of a vast number of legal and financial flows, it, therefore, has a quasi-monopolistic control over information at the national scale, on which it is able to capitalize through its research arm, which will have proven attractive to an economist like Page. PRS 'allows for data sharing for a number of areas in the music business; from sales to live music data to broadcast and online', an aggregational position that, their website claims, allows them 'to shape business development, strategy and planning as well as to inform debate and lead industry thought topics and trends'.[2] Meanwhile, proclaiming itself the 'R&D department for the entire music industry' (Hu 2018), Spotify took advantage of subscriber data for analyses of consumption trends to follow firms like Google (Schmidt and Rosenberg 2014) and package internally developed business fads (such as its agile management model of 'squads', 'chapters', 'tribes' and 'guilds') for broader consumption.[3] One could go on – the claims are as numerous as they are self-congratulatory. The broader point is to map not just a new discursive terrain, but the avenues such organizations and market framings provide for new forms of professional expertise to become established. Importantly, it is an expertise in which different forms of knowledge and training collide, compete and combine, rather than any kind of disciplinary purity.

Disciplining the music business

The various genres of cultural, scholarly and business knowledge in which these protagonists trade – news, market information, popular histories, biographies, academic studies, instruction manuals, self-help guides and data reports – travel with them like luggage through the circuits of musical capitalism. The habit of professing one's expertise was evident through the growing number and visibility of executives and consultants in conference panels, advisory meetings, book contracts, lecture rooms and training workshops. This custom also, as my third list indicates, moved into workplaces. Extended excurses on the precise content and arguments of items on this broad curriculum – their empirical validity or literary value – were uncommon. Their existence and currency around the office did nonetheless indicate something of the self-identity of practitioners as professionals and experts in their field, and equally of the company that employed them as, in Jeff's terminology, a 'centre of excellence' and a space for 'disruptive conversations':

> *I feel, particularly at [Sonaverse], genuinely that it is a centre of excellence for good young people here. Particularly investment in digital. [. . .] The music industry is attracting maybe new kinds of people, from different disciplines [. . .] I look at this floor and I see people whose expertise is in the diversification of a brand, for example. Or how you build a commercial portfolio around an existing audience base. Those are disciplines that aren't traditionally core recorded music disciplines. But those are disciplines that reflect a business that's pivoting around . . . love for music. [. . .] There's definitely an interesting mix and meeting there. I think that's where the business gets really interesting, where you've got . . . that new person coming in who's got a triple-A distinction from Harvard Business School having a productive meeting with somebody who's been signing hits for twenty years. That's really exciting. [. . .] If there are ever moments of conflict or differences of opinion, or a difference of sensibility or ideology, well that's great. I think where you've got a problem is where there's no dissonance whatsoever. That's when the ship just keeps sailing.*

Jeff, interview (2014)

This picture painted here is of the corporation as an agonistic interdisciplinary space. Its occupants are motivated by an idea of progress associated with 'moments of conflict' while consensus is associated with stagnation ('the ship just keeps sailing'). The implicit reference here is to an older industry, portrayed as too bulky and doctrinaire to 'change course' in the face of disruption. The answer, then, is to create 'dissonance', introduce new 'disciplines', allow for clashes of 'ideology' and foster an internal innovation culture in which Thought Leaders and Record Men forge 'disruptive' collaborations. While this collegial imagery is idealized, likely imported from the world of tech, signifiers of academia and professional training circulated widely. These included social sciences. Practitioners evangelized for audience research and talked with sincerity about their responsibility to supply young people with music not only for entertainment but as appropriate resources for identity construction. Likely for my benefit, adroit references were made to Pierre Bourdieu and Max Weber; a vocabulary of 'sub-cultural theory', 'cultural capital', 'dialectics' and 'semiotics'. Bourdieu would surely have appreciated how he himself had become a signifier of intellectual capital in this context of intermediation.

A sense of the 'new kinds of people' identified by Jeff is given by industry analyst Martin. Confessing his professional naïvety when first joining the industry, he moves to both emphasize his formal disciplinary training and his distance from the academic sphere:

I did a degree in economics. A job came up at [music organisation] as an economist, I thought 'OK, that could be really interesting'. Didn't understand the music industry at all – so it's interesting to talk about skills coming out of university! I got the interview based on the economics alone. I knew nothing. Couldn't tell you the distinction between [a label and a collection society], even at the time of the interview. [. . .] I learned the industry with an economic slant [. . .] very much in the way my degree had been taught: cause/effect; isolate impacts [. . .] In that regard, the academic applied very nicely to the real-world situation. If anything though – trying to think of the right way to put it but – the complexity that you see from the inside actually exceeds what the academics see. So, a lot of academics will comment on the music industry without understanding the complexity that exists when there are two rights in a song. And it's quite incredible to watch how theory is applied badly to the real world – rather than watching the real world and seeing theory in action. If you appreciate the distinction.

Martin, interview (2014)

Martin's professed ability to both 'isolate impacts' and to appreciate 'complexity' positions him above those scholarly outsiders who defer to 'theory', failing to understand 'the real world'. Selling himself in his new career as a freelance consultant, my conversation with Martin saw him continually walking this tightrope between the authority of expertise and the humility of professional experience. 'If you're trying to write a hit song', he opined, 'me and a spreadsheet are not gonna help [. . .] with all the talk of data – we risk losing the fact that the creative spark actually makes it what it is'. What is needed instead is a more subtle grasp of 'theory in action'.

Training up

With this in mind, I turn finally to the contested issue of professional development. Interviewees were typically better predisposed towards Sonaverse's in-house education programmes than, as we saw in the previous chapter, the idea of formal academic training. This scepticism was captured most eloquently by Andrew:

> *This is not a business that- It's not a hospital. It is not an architect's practice or an engineering firm. It's not an organisation whereby the individuals need a meticulous understanding of a clinical procedure. It's not an organisation where employees need a rigorous understanding of a mechanical or architectural process. You know, no-one here needs to understand what a load-bearing tangent is; no-one here needs to understand how to do a suture. This is [. . .] a human business. And within two months of graduating, I can be going through long-form license agreements and contracts – I might not know exactly what they are but I can read, I can teach myself what they say and mean – but I don't need to be a lawyer to do that.*
>
> Andrew, interview (2014)

In contrast to the 'triple-A distinction from Harvard Business School', the notion of teaching oneself and learning by doing remained a powerful refrain. As such, Andrew characterized 'big data' as a 'fashionable business discipline' – a mere 'accoutrement' and 'something that people are supposed to be doing now'. In contrast to the professional disciplines of architecture, engineering, medicine, law or indeed data analytics, Andrew emphasized the importance of cultural expertise and an intangible sense of 'good':

> *You don't need to have a degree in music, in terms of composition and orchestration, to know what a good tune sounds like. [. . .] OK, it might benefit to have spent a few hours in a record store, and a few years listening to good records, and to*

be surrounded by people that know good music. I mean, yeah, you know, that's education! That's helpful. But it's not about just being set up to do tasks.

Andrew, interview (2014)

Formal knowledge acquisition is downplayed here in favour of a civilizing vision of culture as the formation of appropriate forms of taste, aesthetic judgement and a network of contacts. Bourdieu, once again, lurks in the background.

Assuming that new recruits arrived with the appropriate level of cultural formation, Andrew admitted that their skills could often be usefully supplemented with generic technical and office skills. He claimed to 'sweat the assets that I've got', in the form of Sonaverse's 'suite of courses [. . .] from financial aptitude to crisis management'. This 'corporate workbook' enabled those under him to develop 'core project management skills, delegation, time management' – all, in his assessment, simply 'basic life skills'. Nonetheless, he admitted, 'in terms of efficacy, they really, really vary' and he warned of the dangers of buying into an accounting firm mentality. Perhaps with EMI's fate in mind, he recoiled from those 'creatively inert' consultants, immersed in the likes of 'Prince2 methodology', whose decision-making – 'completely dictated by technical protocols and Microsoft' – he saw as a 'complete anathema to the music industry'.[4] Andrew's concerns spoke to a (lost) historical sense, as Tony commented with some melancholy (in Chapter 3), of businesses that had a duty 'to train people in the *music industry*': to offer, as the latter considered EMI once did, a holistic 'apprenticeship' in care and custodianship for music rather than skills training for a particular occupation or task.

Nonetheless, Sonaverse's employees did regularly speak of, and I participated in, initiatives to support a general familiarity with the particular structures, contracts, finances, royalty flows (and so on) of this industry. Typically, these were short courses, lasting a day or less, delivered by senior executives who discussed case studies drawn from their own practice. Some of these were descriptive and explanatory ('How it Works'); others were more experiential and task-oriented ('How To'); all were littered with stories and name-dropping anecdotes drawn from a life of experience, adding flesh and colour to the paperwork. Generally, such courses were populated by neophytes like myself, many of whom had never formally studied the industry, alongside those who had recently joined from another industry at middle-management level. While designed to orient employees to the contexts of their work, they also encouraged the kind of 'cross-disciplinary' internal exchange and networking that Jeff advocated.

In addition to the industry explainers and the corporate workbook, there was a set of courses derived from the disciplines of human resources and organizational psychology. I attended a number of these: typically longer, or multi-part, and experiential, involving substantial role-play and confessional exercises. Commonly, for example, we were handed psychometric tests to diagnose Jungian archetypes (hero, caregiver, creator, outlaw, magician, etc.) or Myers-Briggs personality typologies (introversion/extraversion, sensing/ intuition, thinking/feeling, judging/perceiving). These would then be tested out in partners, provoking reflection on our working identities, the traits we project and those we perceive as employees, colleagues and professionals. The 'transactional analysis' framework of Parent-Adult-Child (Harris 1969) might also be mobilized, encouraging us to notice the relations of micro-power in our interactions with others – to diagnose dynamics of domination or support, praise-seeking or scolding and so on – and adjust our behaviour accordingly to achieve the most favourable results. These kinds of '"soft", "cultural" technologies', to use Du Gay's (1996: 142) Foucault-influenced language, have long been part of the managerial toolkit, especially for those in customer service, said to equip the workforce with skills of observation and reflection, 'adding value both to themselves and the company', and empowering them as entrepreneurs of their own fate. Within Sonaverse, this was deemed especially important both when dealing with senior executives – so-called 'managing upwards' – and with external clients (artists, competitors or commercial partners).

While some of these courses offered only a brief diversionary respite from the normal working day, others were more intensive. In the summer of 2010, I attended a five-day residential workshop on strategic negotiation. This form of immersion, common to business education, promises to provide access to what Fabian Muniesa (2017), in reference to Harvard Business School's experiential case-based approach to pedagogy, has described as a 'business real': simulating the pressures of decision-making in conditions of complexity, uncertainty and accountability, often prompting encounters with deeper regions of the self, including anxiety and trauma. Accordingly, working hours ran from an 8.00 am breakfast through to late-night conversation, with no 'off' time. The atmosphere was immersive and designed to fuel competition; a prize was promised to the highest performer. During this time, participants were expected to engage in rote learning of theory – including principles of game theory and transactional analysis – alongside extensive role-playing and behavioural reflection. We were encouraged to internalize a sense of our actions as moves in a zero-sum game (always either winner or loser) since, we were informed, 'collaborative bargaining' has a minimal presence in commercial negotiations. To explore this, we were supplied with a handbook containing a series of models and case studies from management literature. We then practised our tactical and argumentative skills against one another, breaking to pass comments on the language, tone of voice and posture that were deployed. Some of these were filmed: watching back the replay multiple times, the instructor zoned in on a particular gesture or phrase: rewind, repeat, pause, 'What happened there? how did it make you feel? who's got the power?'; rewind, repeat, pause, 'Do you see now? will you do that again?' Brutally honest feedback towards one another's performance was encouraged. The deal was on the table, and if we wanted it, then we would need to take control by setting the terms of inquiry, hypothesizing, paying close attention, collecting data, deducing intentions behind acts, exposing inconsistency and falsity. Theory in action.

At the core of the handbook was a quotation from 'Francis Bacon, Statesman & Philosopher', expressed as something of a mantra: 'Listen, not to contradict and dispute, nor to believe and take for granted, but to weigh and consider'.[5] This captures the kind of close, active listening that we were being taught to hone – not so far removed from that described in Chapter 5 – listening with the intent to form a judgement. But, whereas music might normally be located in a 'passional regime', compelling us to submit to the intense emotional fever of action and engagement, here the express purpose was to fashion a disposition aligned with the more 'paranoid regime' of science: searching, relentlessly, to unveil the hidden explanatory truth lying beneath (Deleuze and Guattari 2004: 123–64). On the morning of day one, each participant was handed a spiral-bound notebook, on the first page of which we were instructed to inscribe, in block capitals:

THEY ONLY LISTEN TO FIND OUT WHERE YOU ARE WRONG.
THEY ONLY LISTEN TO USE IT AGAINST YOU.
NEGOTIATION IS SILENCE. NEGOTIATION IS LISTENING.

We were encouraged, then, to develop a practice of 'symptomatic listening' – *sadistic* even, as cultural theorist Steven Connor (2014) puts it – that 'listens out, not just for what you say, but for how you say it': as when the psychoanalyst's exacting ear trains the patient to listen to, and interpret, themselves. Cast as behavioural theorists, simultaneously of ourselves and each other, we became both subjects and objects of continuous interrogation. In this way, the negotiations course attuned us to forensically monitor our physical deportment as much as our turn of phrase, in an exercise of judicious gathering and calculative release of information.

As for me, what effects did this experience have on my own professional practice and identity? Reflecting back over a decade later, my recollection still feels visceral and intense, even if the description somehow reads like satire. Yet, if I am honest, such a high degree of sustained commentary on my performance and focused self-interrogation, entirely new to me at that time, was indeed transformative. I was certainly far more able to *make sense* of commercial decision-making, both as a generic practice and in relation to music. Moreover, against my own expectations, I found myself developing a *taste* for getting a good deal. Whether I became a more effective negotiator is debatable. It was made clear to me, throughout and in a low-scoring final assessment, that my decision-making ultimately suffered from what the instructor called 'irrational' behaviour: kindly, we could call this altruism; more simply, it was a failure in the moment to understand the rules of the game and so to probabilistically calculate its outcomes. I did not win the prize – a rather lame plastic trophy – which had, in any case, been stolen by one impish A&R, who had had quite enough of being told what to do (our red-faced instructor, much to his annoyance, had not accounted for this particular negotiation tactic and threatened to take the matter to our superiors). My own response was to embrace the failure, resurfacing a latent belief that this was not the environment for me. Failing to adequately perform, and in turn being publicly berated for a lack of profit-making nous, vindicated my 'critical' inclinations and reaffirmed for me a strong sense of the boundary separating the inside from the outside of this community of practice. Likely, it catalysed my retreat – and ultimate exit – into a more 'academic' persona, perhaps one where commercial failure might somehow be repurposed as a badge of pride.

Conclusion

Everyone's a critic

The master's house

Outside the members' club we do the usual post-interview small talk but, just before parting ways, the tone changes. 'And now for the selfish question: if I'm looking for some extra hands at any point, how are you fixed to help out?' Selfish? My heart leapt! Money is tight and, with a baby on the way, my career prospects play regularly on my mind. It's not clear this academia thing's gonna pay off – but what are my options? I probably need to be building experience elsewhere. And yet, I'm not sure that I do want to 'help out'. I try to strike a balance between enthusiasm and non-committal. 'I'm pretty jam-packed right now, I'll likely be looking for opportunities in a few months'. Opportunity. I need to be opportunistic. The feeling of research-as-networking has become familiar but the idea of 'leveraging' this particular contact for work hadn't occurred to me until now. Lack of confidence? But then it's not networking if it feels like networking – or so they say. I recall a misunderstanding over breakfast, when my partner had responded to my mention of 'the interview this afternoon' with a quizzical, 'Oh – what job are you going for?' The mix-up strikes me now as either ironic or prophetic as I reflect on the conversation I've just held, wondering whether I've been interviewer or interviewee.
                                                                    Diary extract (July 2014)

Back at the end of the 1990s, the popular music scholar and Trotskyist historian Dave Harker produced a scathing commentary on Keith Negus' (1992) analysis of UK record labels. Wary of 'economic reductionism', the latter had retreated instead into an 'individualistic . . . postmodernism' characterized by the 'exclusion (or at best marginalisation) of most macro-political and macro-economic questions, a practice which corresponds remarkably closely, and rather worryingly, to the current ideals of Capital'. Despite, that is, the careful description of intricate 'webs' – the various networks spanning and binding cultural-commercial relations – it seems 'there are no *spiders*' (Harker 1997: 49–50). In the pursuit of ethnographic proximity over critical distance, he warned, academics risk falling into the same impoverished relationship of industrial symbiosis that besets trade journalists. This is a slippery slope. Perhaps, mused Harker, 'the music-entertainment conglomerates of the next millennium will have their own "house-academic", much as 1960s and early 1970s West Coast record companies had their "house-hippies"', whose job it would be to transport and translate critical-theoretical sensibilities inside the walls of the capitalist corporation (Harker 1997: 50). He hoped it was not to be.

At the time of writing, Harker would perhaps have been unaware of subcultural theorist Sarah Thornton's efforts, that same year, to trade on her 'club culture' credentials with a new career in advertising. Swapping Hebdige and Bourdieu for 'business books and trade magazines', and immersing herself quasi-ethnographically in impenetrable acronyms and insider distinctions, she developed 'a reasonable patter which involved the general substitution of words like "culture" and "society" with "market" and "consumer" (e.g. swapping "youth culture" for "youth market", "subculture" for "niche market", "social group" for "consumer group")' – eventually landing a job at a 'bog standard multinational' ad agency as a planner: one of 'the intellectuals of the industry' (Thornton 2003: 60–2). Rather more strangely, Harker did not mention the industry trajectories of established PMS scholars, critics and self-described (if unorthodox) Marxist empiricists: Simon Frith, 'trusted by the music industry's trade organization, the BPI, to be the "independent" chair of the Mercury Music Prize' (Frith 2019: 154); or Dave Laing, former editor of *MusicWeek* and the *Financial Times'* music market data section *Music and Copyright*. Surely, the door to the house had long been open to any academic willing to push against it.

Yet what has happened in the intervening years is perhaps more insidious than multinationals hiring 'trendy academics', as Harker calls them. As documented in previous chapters, major record labels and the 'Big Music' organizational ecosystem of which they are a central component have sought to foster their own in-house intellectuals and curate their own endogenous knowledge systems. Some elements of workforce training can be happily outsourced to the university, in relationships that are increasingly managed through the market discipline of "employability" and professional membership subscriptions (Chapter 8) – but that somewhat fractious relationship can be largely bypassed where needed. Belatedly, and unevenly, industry has entered into academia, not the other way around.

This was, then, the nexus at which I found myself at the end of my interview – unexpectedly, being beckoned by my own reflection. Weighing up what kind of a future I was preparing myself for, I learned that, just by making contact, asking questions and jousting with concepts, I had unwittingly been pitching for work: trading not stories and anecdotes for 'career capital', as John Caldwell (2008) writes, but theories and interpretations. Seeking legitimacy, I was dimly aware of engaging in a performance of intellectual capital. Yet I had not expected the critical ideas and methods in which I was most interested to have such a good exchange rate with their more market-facing equivalents, in the way Sarah Thornton describes. For someone like Dave Harker, this is a compromise too far, on the way to becoming a useful corporate idiot. Intended as an amusingly absurd comparison, the etymology of the 'house hippy' has a darker history, implying, via references to chattel slavery, a willing servility within the master's house.[1] But, as Devon Powers (2012) has documented, it is easy to reduce such figures to a 'type', understating the considerable freedom that such individuals genuinely did have to effect internal change with real positive effects, whether getting interesting and unusual music in front of a mass audience or simply funnelling resources to small radical projects below the radar (during the cash-rich good times, at least). Equally, it overstates their perceived distance from the existing corporate culture in the first place. Likewise, rather than standing on the outside looking in, what my interviewee had identified, which I had likely disavowed, was that we were part of a shared professional world (cf. Ortner 2010). In retrospect, as an over-educated, able-bodied white cis man of a certain age, the invitation into 'the club' (see Chapter 3) was hardly a surprise.

To describe the aggregated organization I called Sonaverse Records, in this book I have embraced the principle of symmetry, following the actors themselves and taking their views seriously, seeking 'critical proximity, not critical distance' (Latour 2005: 253). Whenever I have read a novel, hung out in a post room, watched a reality TV show, stayed late in the office, decorated a desk, taken an Addison Lee to the K West hotel, emailed an artist's manager, shared a Spotify playlist, got my head around a software system, requested an archive box, raised a UPC, engaged with a trade body, subscribed to a mailing list, messaged someone on LinkedIn, spoken at an industry conference or attended a training course, I have done so not to 'lift the lid' on a secretive industry and reveal its inner workings but because I have in some way been prompted to do so, whether by research participants, colleagues or managers. In such routine ways, the music economy is continuously and actively 'provoked' into being (Muniesa 2014). But this does not mean simply taking people at their word; rather, I have pursued a critical-pragmatic approach that aspires to be 'capable of taking account of the ways in which people engage in action, their justifications, and the meaning they give to their actions' on an everyday basis (Boltanski and Chiapello 2005: 3). Doing so has enabled me to trace a set of relationships at the heart of the global network that sustains a particular mode of production, which I called Big Music. If this hardly exhausts or fully explains the field, it nonetheless enabled me to articulate the myriad ways in which this particular organizational form is brought into being and rendered as legitimate. In the next section, I clarify the twin processes of 'top-down' managerial *passionalization* and the 'bottom-up' *meaningful maintenance* practised by employees in order to highlight the gap that emerges between them and consider what opportunities for intervention arise as a result.

But make no mistake, there are spiders in this web. And so, by way of conclusion, I want to return to the question of critique. If the period analysed in this book saw little in the way of protest or resistance – a silence which I found increasingly troubling during the research phase while also recognizing a degree of my own complicity – the years since have been marked, in part and with good reason, by a veritable flourishing of critical voices. A new spirit is still emerging. The challenge that remains, I suggest, is to find ways of connecting this to the kind of mundane maintenance work that is needed to build new systems and new worlds.

Passionalization and meaningful maintenance

Across the book, I have been interested in the internal legitimacy of industrial change, noting two tendencies. On the one hand, the abstraction of corporate bureaucracy: rationalizing processes of streamlining and tightening integration between legal frameworks, technical protocols, managerial procedures and statistical classifications; that is, the abstraction of corporate bureaucracy. On the other hand, a more singular 'cultural' domain, captured by the common-sense understanding of the 'passion for music': the circulation of ethical values, affective attachments and interesting ideas in social worlds. While distinct, I have argued that the two tendencies are entwined and mutually reinforcing.

And so music work has been represented, read, recommended and interpreted in various forms, alongside an ill-defined and shifting curriculum of popular business literature. These circulate a loose anthropology of music engagement along an active-passive spectrum, connecting personal stories to a market imaginary that segments and quantifies the 'consumer' for differentiated product campaigns. Music workers' self-understanding partially aligns with algorithmic recommendation functions in such ways. Sonaverse has equally driven forward the implementation of International Standard Recording Codes (among other information standards), alongside greater agreement among policy communities over copyright regulation, which have together stabilized flows of intangible data, enabling them to be tracked through a dislocated production network. Their global enterprise systems shape the coordination of those parts of the value chain housed within the corporation, as well as the many external suppliers that have to deal with it, for purposes of financial and metadata tracking. The employees who generate and execute these information flows meanwhile participate in a community of style and taste, communicated through clothing, musical preferences, office soundtracks and gig attendance, which is then recoded through managerial instructions and public comms initiatives that convey a suitable internal 'atmosphere'. Employees are able to position themselves within a broader imagined community, assembled through curated lists of news items, executive interviews and insider gossip, through which particular business dialects and norms can be learned. Efforts to standardize and 'kitemark' a certain level or quality of music business education, in order to train up and professionalize the workforce, are ongoing. But the originary myth of the record men flows through all of these spaces, handed down as oral culture.

These processes are largely sanctioned by Sonaverse, top-down, connecting the supply chains and music cultures outside the organization with the workforce inside, in a hybrid form of what I am calling 'passionalization': the drive to rationalize, increase efficiency and agility at scale, undergirded and legitimated by appeals to the 'passion for music'. Thus, the company could be strategically remade in ways that profitably involve, engage and deploy its workforce, with the specific set of commitments, ideals and expertise they bring, but not always to their advantage. Such initiatives have produced uneven successes; many were delivered clumsily and stiltingly by people not quite ready to put their hearts into it; some were contested, later to be swiftly rolled back. But in order to take place at all, the organization needs simply to *go on* – and it does so thanks to people performing not groundbreaking acts of innovation but the routines that maintain working life: emailing spreadsheets, inputting data, logging calls, looking up information, holding meetings, filing contracts and so on. Still, they are largely able to express a sense that this work is meaningful – that it is performed alongside other interesting people; that it is interesting insofar as it doesn't follow the same rules and codes of conduct as everywhere else; that it contributes to something important and bigger than them – even while identifying themselves, on occasion, with technical procedure and organizational protections. They recognize the cultural needs of the organization while disavowing their own tastes in favour of 'what works' for artists and fans. As such, they are not just finding meaning in the mundane; they are engaged in *maintaining meaning*: sustaining the conditions in which they and others can locate themselves as people who care about music.

Remaking the major label has relied, then, on establishing a 'new spirit' that can sustain and give purpose to work. Stories of excess and transgression continued to circulate, along with many of the individuals that populated those stories. But the ethical concern to connect and support the artist-fan relationship was genuine and palpable, whether in interviews or off-the-record comments and asides. Sonaverse's capacity to appeal to and channel that concern was crucial to its own survival. These employees are not reducible to the organization that employs them. They are thoughtful and reflective – often they are home-grown industry intellectuals, just as Debbie commented in Chapter 3: 'really smart', knowledgeable, articulate and certainly passionate about their craft.

Still others, perhaps a majority, are cynical about their employer's motives, seeing the job as a short-term compromise that pays the bills and looks good on the CV, or have resigned themselves to a career they are no longer sure they want to be in – but they too need to be 'brought along'. After a few drinks at an industry event, they prove painfully aware of the popular stereotype they are performing: of being duped by music biz 'glamour', sated with the bread and circuses of occasional gig tickets or awards ceremonies. But few employees expend much energy worrying what their industry leaders and trade representatives are up to, aside from a suspicion that they might be about to interfere with their work. Many are likely to agree with one individual's assessment of a senior exec: just 'a nob with big words' who, instead of making managerial decisions based on 'what the research says', should just go out and 'sell some records'. In short, the company is inhabited with complex, multi-faceted humans whose cares and curiosities are hardly exhausted by relentless, mechanical appeals to a 'passion for music' – but equally whose energies are not unlimited. Within this context, it is quite plausible that the employees of major labels will 'find their everyday universe uninhabitable' (Boltanski and Chiapello 2005: 10), that their workplace will provoke private indignation that connects to public critique.

Swift exits, muted voices, tested loyalties

Boltanski and Chiapello build their understanding of critique on a framework from economic historian Albert Hirschmann, whose description of 'the passions and the interests' we met in Chapter 4, but who also had a thing or two to say about dissatisfaction in workplaces, consumer markets and nations. For Hirschmann (1970), when people are aggrieved in liberal market societies, they are presented with the option to 'exit': simply to leave and take their money, their labour power or their citizenship elsewhere. Yet many choose instead to exercise their 'voice', raising their concerns in complaints or critiques of the issues at stake in the hope of change. 'Exit' is the market option, while 'Voice' is the political option. Hirschmann's insight is that, while the critical voices can be difficult to listen to, they are usually predicated on what he terms 'Loyalty' and the will to improve situations. *Criticism* is an expression, albeit a contingent one, of *commitment*.

During the majority of the period under study, Sonaverse's most vocal critics were those largely outside the organization with very little in the way of loyalty. Meanwhile, its employees reflected on their own employment conditions only rarely. In interviews, many seemed to have difficulty grasping or accepting that their work might be of research significance, so that responses would frequently default to talk about 'artists' or 'the industry' more generally rather than the specific tasks and conditions of their own employment. The language did not seem available to them – unsurprisingly, given that there is very little collective labour representation available either, only HR. On enquiry, neither of the likely suspects, the Musicians' Union (MU) nor Broadcasting, Entertainment, Communications and Theatre Union (BECTU), would accept record company staff as members.[2] Hence, the employee's 'voice' is relatively muted. According to 2011 statistics, over half of workers in recorded music left the sector within five years and over 75 per cent within ten (CC Skills 2011: 18): *exit*, voluntary or involuntary, seemed to be the primary option. Despite, or perhaps because of, the pervasive threat of redundancy hanging over them, as Mike Jones (2012: 115) notes, back-office staff 'seem stoically to accept such practices; there have been no strikes by music company personnel in music industry history', notwithstanding the female packing workers discussed in Chapter 1. Indeed, in line with processes of 'passionalization', they are themselves liable to justify and rationalize cutbacks, not by way of reference to the calculative cost-benefit analyses of austere times but, as Sophie described (Chapter 3), to an internal culture that needs to be periodically 'refreshed', and its spirit renewed with younger voices. Given this, relatively minor disputes such as that over the office soundtrack (Chapter 5) appear far more significant as an emergent means of giving voice to felt grievances in ways that are intrinsic to the craft of maintaining meaning.

It is perhaps for such reasons, allied to the dominance of the 'artistic critique' and a long-standing spirit of hedonism and transgression, that a range of inequalities seemed in some way 'unspeakable' – jarring with a meritocratic self-image (Gill 2014; Taylor and O'Brien 2017). Employees appeared poorly attuned to a range of social injustices within their workplace, albeit to different degrees and in different capacities. From 2013, though, something seemed to change. The appearance of a number of sexually explicit music videos – most prominently Miley Cyrus' 'Wrecking Ball' and 'Blurred Lines' by Robin Thicke and Pharrell – buoyed a number of public debates that soon moved inside Sonaverse (for an extended discussion of the issues raised in this section, see Bennett 2018b). 'Blurred Lines' in particular paraded lyrics that appeared to sanction sexual harassment and potentially rape. Banned by student unions and spurring government proposals to subject music videos to age ratings, media reporting around the role of industry in propagating misogyny soon moved to discussions of a 'behind the scenes [. . .] gender imbalance' in work (CMU Editorial 2014), dovetailing neatly with another, more established, debate around internships, recruitment practices and uneven pathways into creative work (see Chapter 2).

In this context of rising public critique, the absence of direct commentary from interviewees became increasingly troubling. They might designate certain individuals or actions as sexist, ageist or otherwise unjustly exclusionary in informal conversation – but often such instances were raised as an apparently neutral description of the workplace rather than a complaint, and usually when the voice recorder was switched off. Comms executive Alan was the exception when, in a comment on what he dismissively called 'the whole feminist taking-your-clothes-off thing', he advised me to avoid this territory: 'please don't follow that one, it's slightly . . . I'm sure a lot of people have already written on it!' Clearly, my interviewees were not mere data sources. They had opinions on what was and was not (should and should not be) of academic interest. I was encouraged to produce my own silences.

Alan's comments, coupled with his professed 'protective' stance towards his colleagues and the wider industry (Chapter 5 again), provoked me to reflect further on the world in which I had been immersed – and indeed passively complicit – and the anecdotes and asides that had piled up. The veteran A&R,

a recognized 'eccentric' with a darkened office referred to as 'the womb', caught smelling the seat of the female artist who'd just vacated it. The junior marketing assistant walking into an office of white faces each day and feeling 'This is not where I belong'. The venue owner, who'd had a VIP room installed *behind* the VIP room in order to receive audiences with 'pretty young girls'. The reference to the newly signed group of working-class boys as 'a bunch of car thieves from Bolton'. The repeated mentions of individuals 'playing the race card'. The young employee who paired the predatory 'seduction' techniques of 'pick-up artists', such as 'negging' (insulting or ignoring) female 'targets', with carefully selected namedrops to his rock royalty connections. The interviewee who requested not to mention their gender or ethnicity alongside their profession, as this would effectively de-anonymize their testimony. The one Black member of the team who was consistently asked for an opinion on new R&B releases – a genre he never listened to. The project manager who accompanied a female counterpart to a product meeting with the words, 'shall I bring lube – or is saliva OK?' And so on, and so on. It is less the accuracy of such tales that should give pause here, than their discursive currency – wheeled out and regaled to an audience, sometimes by the very people they were about, with a resigned shrug and a 'well, that's the music business!'

The critical age

Within the discursive limits of this production culture, the 'feminist taking-your-clothes-off thing' marked out a rare critical space. Individuals and trade commentators were enabled to voice concerns over topics previously considered sensitive, controversial or even 'unspeakable' (Gill 2014), while a host of Equality, Diversity and Inclusion (EDI) initiatives, reports and support networks were set up (Bennett 2018b). This came to a head in 2017, when the publicity around a culture of sexual harassment within Hollywood, generated by the allegations faced by Miramax producer Harvey Weinstein and the far broader #metoo campaign, spread to other industries, including music. A short documentary, aired on BBC television, exposed a similarly 'endemic' atmosphere of 'sexual assault and abuse' (Mackenzie 2017). UK Music, the sector's lobbying body, established a Diversity Task Force and Equality and Diversity Charter, alongside their Meritocracy Dinner Series. Their most significant intervention was a 'diversity survey' – a data-collection exercise promoted through member organizations and trade press – significant both as a serious attempt to address these issues with new statistics and because its results were then included in the body's economic estimates, which are used by the government (UK Music 2017: 17).

A second wave of critique arrived in music in 2020, during the Covid-19 pandemic, after the murder of George Floyd and the subsequent rise in global visibility of the Black Lives Matter (BLM) movement. This time, institutions, especially major entertainment companies, were much faster to respond, starting with a social media 'blackout' gesture of solidarity, replacing company logos with black squares. Internally, a series of organizational decisions were also made, perhaps best captured in Universal's tellingly-named Global Taskforce for Meaningful Change (implying that previous initiatives had been rather less so) and the independent Black Music Coalition, which sought to connect internal changes around policies and procedures to the financing of education initiatives and charitable support for housing and legal aid. Such coalitions led to, for example, the dissolution of 'urban' as an organizational division and the cancellation of unpaid royalties for heritage acts that had yet to recoup their advances, in which Black artists were over-represented (Ingham 2021).

A concurrent but largely separate set of critiques has concerned the rise of streaming platforms (Hesmondhalgh 2022; Marshall 2015). Internal negotiations between platforms and rightsholders concern the correct balance of remuneration rates, exceptions and subscription pricing. A public debate, meanwhile, has revolved around, first, the social questions of just reward and the extraction of value from artists and songwriters, on one side, and platform users (in the form of personal and behavioural data collection) on the other. And second, an artistic critique persists in accusations that platforms reduce music to its utility function: transforming the dedication of artistic craft into an abundance of passive, undifferentiated wallpaper moods for tuning the focus and productivity of students and knowledge workers (major label employees, for instance), deadening listeners' aesthetic and critical faculties. In the UK, the #BrokenRecord campaign took the lead here, spearheaded by Tom Gray of the Britpop-era band Gomez, pitched at the popular press, industry leaders and policymakers, supported by a range of bodies, including the Musicians' Union. This led to a widely publicized government inquiry in 2020 and a commitment on the part of some majors to explore alternative models (CMU n.d.).

These critiques are novel and important in a range of ways, not least because they often transcend normative divisions between creative 'talent' and the non-creative workforce, suggesting new configurations. The public response to music videos in 2013 raised a broader discussion 'behind the scenes', blurring lines (so to speak) between music's reception and its everyday reproduction. BLM allowed a conversation in which issues of rights and royalties were shown to be rooted in toxic workplaces. In other words, rather than simply rebalancing numbers, they demonstrated the fusion of the production system with its production culture: both *how it works* and *why it matters*. MeToo and BLM coupled, to use the political theorist Nancy Fraser's (2013) language, a politics of 'recognition' (who counts and in what way?) to a politics of 'redistribution' (who benefits and who pays?), and so began to raise a hitherto absent politics of representation (who's visible, who's audible?), in the form of a collective labouring identity. In the United States, the United Musicians and Allied Workers campaign group was formed in 2020, prompting a more formal unionization drive at the Secretly Group of independent record labels, ratified in 2022, with Bandcamp United recognized in 2023. Significantly, both are allied not with performing and songwriting organizations but to the Office and Professional Employees' International Union. They buck the broader trend and mirror – albeit to a much lesser and more fractious degree – emerging forms of workplace organizing in new journalism and technology companies (Cohen and de Peuter 2020; Tarnoff and Weigel 2020). Notably, while expressly representing 'non-creative' administrative employees, they stress connections with creative communities: 'Many of us work at Bandcamp because we agree with the values the company upholds for artists [. . .] We have organized a union to ensure that Bandcamp treats their workers with these same values' (Kelly 2023). While unlikely, it is not inconceivable that a major label union could follow. This in itself is significant: the boundaries around what is thinkable have shifted considerably.

Exploitation (in the nicest possible way)

Recorded music is still about taking the content, the music that we have, and in one way or the other, trying to create demand, and then exploiting that demand – 'exploiting' in the nicest possible way . . .!

Jeff, interview (2014)

Few of the aforementioned critiques, however, address the contradiction at the core of this book: the *exploitation of copyrights* that the digital production system was built to support and enhance; the *exploitation of labour* that underpins and reproduces it. When Prince claimed that 'if you don't own your masters, your master owns you', he highlighted the persistence of historic injustices while hardly contesting individual ownership itself (Vats 2019). The 'moral economy' of the rentiership model – of perpetual rights exploitation supported by legislation; realized through a complex technical architecture; coupled to and legitimated by appeals to the passion for music – persists. Culturally, the notion that such a system exists to serve the relationship between artists and fans forcibly occludes all the non-creative maintenance work that goes on in the middle, which has only recently become visible.

Such workers are incredibly productive. In 2022, employees of corporate music divisions generated, on average, just under $1,000,000 of revenue per head (Ingham 2023a) – a figure that is reflected (and more) in the salaries of only a handful. In part, this is explained by efficiency gains thanks to technology investments of the kind discussed over the course of this book; mainly, it is explained by the continued ownership and aggressive exploitation of historic rights catalogues through the work carried out by such employees. The current wave of apparently disruptive innovations, in the form of a variety of blockchain and so-called artificial intelligence technologies, is not in principle incompatible with this process of passionalization: the one simply claiming, once again, to do away with unnecessary administrative intermediation in the name of enabling artists; the other relying on access to copyrighted content in the form of large training datasets, the most valuable of which belong to major rightsholders (who are, at the time of writing, lobbying aggressively while negotiating new licenses). As such, by the opening weeks of 2024 the promise of mass redundancy loomed once more, in what record industry leaders described as a new 'cut to grow' phase, with a continued emphasis on catalogue placing pressures on frontline back office staff (Ingham 2023b). But the need for maintenance does not disappear, and in order for it to be effective, it must be meaningful.

Maintaining musical meaning at scale is a worthy goal – but this does not in itself require the extractive exploitation of people and assets by corporate oligopolies. If new systems are to arise, ones that would break with the passionalized rentier model of tightly controlled ownership, they will, of course, need significant quantities of material investment; they will need to work with and intervene in global policy and regulatory frameworks; but they will also require collective labour. That will demand a new spirit to which everyday routine workers, as well as ideological true believers, can commit. The back-office admins I spoke with repeatedly stressed occupational humility within the context of a broader artistic and cultural community, seeing themselves as united not by shared practices but by a common project. At the same time, they exhibited strong defences of bureaucracy (duty, due process and accountability) which often sparked indignation towards 'creative' colleagues' romanticized misperceptions and misdemeanours.

Conceivably, then, rather than merely following orders, commercial music admin holds within it an emergent but underarticulated 'ethics of office' (Du Gay 2017) – a more prudent, public-oriented vocation than is commonly mythologized, one embedded within the system. It is the job of any 'critical intellectual', academic or otherwise, not simply to lift the veil and assert the superiority of their own explanations, nor to opportunistically search out opportunities for collaboration and 'impact'. It is to work with and across these internal critiques wherever they arise, connecting the dots, creating space and giving voice to their experiences and concerns, rearticulating them in new forms and forums. To adapt the intersectional feminist refrain, if the 'master's house' of the music corporation is to be 'dismantled', rather than remade once again, then alternative intellectual 'tools' are required (Lorde 2018): explaining *how* the industry it sustains works – but also, crucially, *who* works and *why*. The major label must be dismantled, that is, from the inside out.

Appendices

List of interviewees

Below I list interviewees that spoke with me 'on record'. Interviewees were assigned pseudonyms, or chose their own, and titles that loosely match their job descriptions, erring on the side of caution for anonymity purposes. Most chose not to disclose identifying characteristics.

#	Date of Interview	Name	Employment Description
1	Oct 2013	Olive	. . . an administrator at a major label for around five years, before moving to a managerial position at an independent.
2	Nov 2013	Gavin	. . . a department head at a major label; he has worked for the same company for over twenty years.
3	Jan 2014	Alex	. . . a marketing coordinator whose career has spanned two major labels and an independent.
4	Jan 2014	Alan	. . . worked for several years in a major label communications department, after a brief stint of work experience at a small independent.
5	Jan 2014	Leonard	. . . a veteran of some forty years, recently established as a consultant. Beginning in advertising and music marketing, he has held senior director positions across various companies, including several majors.
6	Feb 2014	Jeff	. . . heads a major label insight department who started life at a brand strategy agency before moving into music.
7	Apr 2014	Ian	. . . a journalist of around ten years' standing, writing predominantly for music and media trade publications.
8	Jul 2014	Martin	. . . an economist by training who had recently set up an independent research agency. He has held strategy positions in two majors.

#	Date of Interview	Name	Employment Description
9	Aug 2014	Melissa	. . . a legal assistant in a core business role, who has worked across rights management and business affairs.
10	Aug 2014	Anonymous	. . . declined to disclose any identifying information.
11	Aug 2014	Sophie	. . . a legal secretary for a frontline label division at a major, who had been at the same company for over a decade.
12	Aug 2014	Andrew	. . . a general manager for a label, with twenty years' experience in multiple marketing and promotions positions across the industry.
13	Aug 2014	Lou	. . . a lecturer on a university creative industries degree programme, specialising in music, with a background as a songwriter and artist manager.
14	Sep 2014	Graham	. . . a lecturer on a music business degree programme. Prior to this he had worked in marketing within major music companies and as an artist manager.
15	Oct 2014	Zoe	. . . works at an independent music publisher, with past experience as talent scout for a major label.
16	Oct 2014	Ben	. . . manages a major-signed band who were starting to gain mainstream recognition. He has a background as a singer-songwriter and in popular music education.
17	Apr 2015	Debbie	. . . spent several years heading a major label insight department, which she left to pursue a market research career.
18	May 2015	Nathan	. . . manages a data administration team at a major record label.
19	May 2015	Tony	. . . general counsel for a major record label, after an established career in music law.
20	Jun 2015	Ash	. . . a product manager for a major label catalogue division, with a background in music television.
21	Jun 2015	Poppy	. . . a digital marketing manager with experience at two major labels.
22	Jun 2015	Rich	. . . a product manager for a major label catalogue division, with a background in audio production, music retail and at an independent label.
23	Jun 2015	Nick	. . . handles business development for a major label. His background lies in strategic planning for a music rights management company and in venture capital.

Corpus of book recommendations

The full corpus of books analysed in Chapter 9 consisted of forty-one in total. They derive from three sources:

Set A (15 titles): Eamonn Forde: 'The 15 Music Industry Books You Must Read', *Music Ally* (Forde 2015)

Set B (25 titles): Will Page: 'Rockonomics', *Amazon Listmania* (Page, n.d.)

Set C (9 titles): Recommendations from interviewees and other research participants.

N.B. Some book choices overlap between sets, so that totals do not sum.

The texts were then coded by genre and subgenre, loosely according to intended audience:

'Textbooks' (19)
- 'How it Works': overviews and explainers (8)
- 'How-to Guides': practical manuals (7)
- 'Academic': scholarly texts (4)
'Popular Nonfiction' (20)
- 'Histories': of companies or industries (17)
- 'Biographies': of individuals (2)
- 'Criticism': a compilation of journalistic writing (1)
'Fiction' (2)
- 'Novel': a dark comic thriller (1)
- 'Film': a dramatization of a theatre play (1)

The texts are dated by the most recent publication date and allocated a 'Theme': an indication of content (whether directly 'about' music business or imported from other contexts). They are presented below in alphabetical order.

Corpus Set	Author	Title	Year	Genre	Subgenre	Theme
B	Anderson, Chris	The Longer Long Tail: How Endless Choice is Creating Unlimited Demand (2nd ed.)	2009	Textbook	How it works	Online Business
C	Bangs, Lester	Psychotic Reactions and Carburettor Dung	1987	Popular	Journalism	Music
B	Brabec, Jeffrey and Todd Brabec	Music, Money and Success: The Insider's Guide to Making Money in the Music Business (7th ed.)	2011	Textbook	How-to Guide	Music Business
B	Budnick, Dean and Josh Baron	Ticket Masters: The Rise of the Concert Industry and How the Public Got Scalped	2011	Popular	History	Music Business
A	Cavanagh, David	The Creation Records Story: My Magpie Eyes Are Hungry For the Prize	2000	Popular	History	Music Company
A	Charnas, Dan	The Big Payback: The History Of The Business of Hip-Hop	2011	Popular	History	Music Business
B	Christensen, Clayton M.	The Innovator's Dilemma: The Revolutionary Book That Will Change the way you do Business	1997	Textbook	How-to Guide	Business Strategy
A	Cornyn, Stan	Exploding: The Highs, Hits, Hype, Heroes & Hustlers Of The Warner Music Group	2002	Popular	History	Music Company
A, B, C	Dannen, Frederick	Hit Men: Power Brokers and Fast Money Inside the Music Business	1990	Popular	History	Music Business
A	Doggett, Peter	You Never Give Me Your Money: The Battle For The Soul Of The Beatles	2009	Popular	History	Artist
B	Elberse, Anita	Blockbusters: Hit-making, Risk-taking and the Big Business of Entertainment	2013	Textbook	How it works	Media Business
B	Eliot, Marc	Rockonomics: The Money Behind the Music	1996	Popular	History	Music Business
B	Goodell Brown, Robert	Statistical Forecasting for Inventory Control	1959	Textbook	Academic	Economics
B	Gordon, Steve	The Future of the Music Business: How to Succeed With the New Digital Technologies. A Guide for Artists and Entrepreneurs (4th ed.)	2015	Textbook	How-to Guide	Music Business

Corpus Set	Author	Title	Year	Genre	Subgenre	Theme
B, C	Harrison, Ann	Music: The Business. The Essential Guide to the Law and the Deals (6th ed.)	2014	Textbook	How-to Guide	Music Business
B	Heller, Michael	Gridlock Economy: How Too Much Ownership Wrecks Markets, Stops Innovation and Costs Lives	2008	Textbook	How it works	Economics
A, C	King, Richard	How Soon Is Now?: The Madmen and Mavericks Who Made Independent Music 1975-2005	2012	Popular	History	Music Business
A, B	Knopper, Steve	Appetite for Self-Destruction: The Spectacular Crash of the Record Industry in the Digital Age	2009	Popular	History	Music Business
C	Kusek, David and Gerd Leonhard	The Future of Music: Manifesto for the Digital Music Revolution	2005	Textbook	How it works	Music Business
C	Loog Oldham, Andrew	Stoned	2001	Popular	Biography	Music Executive
B	Lovell, Nicholas	The Curve. From Freeloaders into Superfans: The Future of Business	2013	Textbook	How-to Guide	Online Business
A	Marks, Craig and Rob Tannenbaum	I Want My MTV: The Uncensored Story Of The Music Video Revolution	2011	Popular	History	Music Company
B	McLuhan, Marshall	Understanding Media: The Extensions of Man	1964	Textbook	Academic	Media Theory
A	Menn, Joseph	All The Rave: The Rise & Fall Of Shawn Fanning's Napster	2003	Popular	History	Online Company
B	Minto, Barbara	The Pyramid Principle: Logic in Writing and Thinking	2008	Textbook	How-to Guide	Business Communication
A	Murphy, Gareth	Cowboys & Indies: The Epic History Of The Record Industry	2014	Popular	History	Music Business
B	N/A	Barbarians at the Gate [DVD]	1993	Fiction	Film	Trading History

	Author	Title	Year			
C	Napier Bell, Simon	Black Vinyl, White Powder	2001	Popular	History	Music Business
A	Napier Bell, Simon	Ta-Ra-Ra-Boom-De-Ay	2014	Popular	History	Music Business
A	Negus, Keith	Producing Pop: Culture and Conflict in the Music Industry	1992	Textbook	Academic	Music Business
B, C	Niven, John	Kill Your Friends	2009	Fiction	Novel	Music Business
B, C	Passman, Donald	All You Need to Know About the Music Business (8th ed.)	2014	Textbook	How-to Guide	Music Business
B	Penenberg, Adam	Viral Loop: The Power of Pass It On	2010	Textbook	How it works	Online Business
B	Poundstone, William	Priceless: The Hidden Psychology of Value	2011	Textbook	How it works	Economics
B	Schwartz, Barry	The Paradox of Choice: Why More is Less	2005	Textbook	How it works	Economics
A	Seabrook, John	The Song Machine: Inside The Hit Factory	2015	Popular	History	Music
A	Southall, Brian	The Rise and Fall of EMI Records	2009	Popular	History	Music Company
B	Suisman, David	Selling Sounds: The Commercial Revolution in American Music	2012	Popular	History	Music Business
B	Szymanski, Stefan	Playbooks and Checkbooks: An Introduction to the Economics of Modern Sports	2009	Textbook	How it works	Sports
B	Varian, Hal M.	Intermediate Macroeconomics: A Modern Approach (8th ed.)	2010	Textbook	Academic	Economics
A, B	Yetnikoff, Walter	Howling at the Moon: The True Story of the Mad Genius of the Music World	2004	Popular	Biography	Music Executive

Notes

Corporate Life in the Digital Music Industry

1　Detailed in Appendix 1.

Chapter 1

1　Such criticism would align with that made by Arts Council England head Alan Davey, who described majors as 'giving the public what they want' (making safe investments) while ACE's role was 'getting the public to find things they didn't know they wanted' (making risky investments). The BPI's Geoff Taylor dismissed the comments as 'ill-informed and out of touch' since 'UK labels have invested £1 billion over the last five years in new music', citing 'Adele, Mumford & Sons, Emeli Sandé, Ed Sheeran, Muse and Jessie J' as evidence of success (Cooke 2013). Whether Davey felt chastened or vindicated is unclear.

Chapter 3

1　'If PIAS damage bad as sounds, implication for UK independent music industry very severe indeed', lamented indie singer-songwriter Chris T-T on Twitter: 'Small dedicated businesses, NOT majors :(' (Vinnicombe 2011). Elsewhere, Oasis' Noel Gallagher offered a sociological judgement: 'The people who are at these riots aren't poor', he said, but 'kids with fucking mobile phones and all sorts of shit', shaped by 'brutal TV and videogames', who had become 'idiots [who] destroy their own communities' (Smirke 2011).

2　Notwithstanding Universal's announcement in 2023 to open a branch office – though notably not its headquarters – of 'EMI North' in the Yorkshire city of Leeds.

Chapter 4

1　Philip Roscoe (2023: 19) considers a similar gonzo-informed approach to 'finding our moral compass in a complicated world' in his depiction of the industry narratives circulating within the world of finance – an industry that came of age at a similar historical moment and would encounter its own legacy of heroic debauchery, under conditions of even more intense pressure and public scrutiny, as well as more ambivalent media representations, only a few years later. I am partially indebted to his discussion here.

2 Reproduced in *Generation of Swine: Tales of Shame and Degradation in the '80s* (1988). Notably, the second edition of the Barfe book corrects the misquote.

3 *Pop Idol* copied and developed the concept of *Popstars* (itself a spinoff of an earlier New Zealand series in 1999), which followed the manufacture of pop group Hear'Say from audition to market, first broadcast in the UK in January that year. However, that show took a documentary format, with less involvement of the home audience. Its US equivalent, *American Idol*, was launched in 2002 with Cowell as judge, initiating a fifteen-year run.

4 This particular reference is overcoded with racialized and classed associations: 'bling' being the term popularized through hip-hop culture to capture the 'sound' of garish, gleaming jewellery.

Chapter 5

1 Extracts – in reality, these are far more extensive.

2 Again, this is a fictionalized impression of a typical slogan.

3 The company Graze offered subscribers a regular delivery of healthy snack foods for desk-bound office workers, as an alternative to absent-minded snacking. Bootcamp fitness – a short and intense open-air exercise regime in the style of military training – was a popular lunchtime activity.

4 This is derived from the fanBase website (fanBase n.d.). The process of developing and adopting this model at Sony is explained – and self-mythologized – by the Colourtext marketing agency, founded by former Emap researcher Jason Brownlee: https://www.youtube.com/watch?v=crH1zJV-e8Q

5 While Ian is not a Sonaverse employee, his particularly articulate account is unexceptional within the major label context.

6 This conversation is once more indicative, impressionistically assembled from many overheard in various contexts.

Chapter 6

1 Millward Brown and Official Charts Company are market research and sales tracking firms, maintaining various crucial rankings and charts. MCPS (part of PRS

for Music) and PPL are collective management organizations that administer royalty payments to rights owners.

2   All quotes in this paragraph and the next from Cornine (Consider Solutions 2016).

Chapter 7

1   One of this book's early readers expressed scepticism here. To take two examples, blockchain technologies (especially 'smart contracts') aim to disintermediate administration by automating trust and transparency, while artificial intelligence tools aim to generate sophisticated and creative responses to human prompts. Both of these produce efficiencies and expand productive capacity. Yet the history of workplace mechanization and automation, as well as that of domestic labour, suggests that such changes will not do away with routine; instead, we typically witness deskilling, displacement and devaluation within organizations, outsourcing of certain tasks to everyday consumers and system users, and new forms of systems management. Such processes are not dissimilar to those described in this chapter. Neither blockchain nor AI can operate without human input somewhere, in the form of training datasets, checking and oversight monitoring, system maintenance and so on. New forms of 'back office' will likely emerge. But the argument of previous chapters has been to stress the value of relationships, tacit understanding and emotional labour that mediate music's various systems, creatives and executive decision-makers, suggesting a continuing need for proximity too. Ultimately, this book is not in the business of forecasting. But if its overall argument holds any weight, it warrants scepticism towards the idea that such technologies have relegated the need for human administration to the past.

Chapter 8

1   The education entrepreneur Norton York is a key figure here, having been central to both RockSchool certification and, alongside the late PMS academic Dave Laing, the establishment of Westminster's commercial music BA.

Chapter 9

1   http://musically.com/ Accessed 24 November 2023.

2   https://www.prsformusic.com/what-we-do/influencing-policy/research Accessed 24 November 2023.

3   https://engineering.atspotify.com/2014/03/spotify-engineering-culture-part-1/ Accessed 24 November 2023

4   Prince2 is a strict rules-based project management technique that has become an industry and government standard, under the banner of which exists a sub-industry of training courses and qualifications.

5   Like the Hunter S. Thompson epithet discussed in Chapter 4, this is another misquote: tellingly, perhaps, Bacon was referring to *reading* rather than listening.

Conclusion

1   Described in a 1968 *Rolling Stone* article as 'the record company's equivalent of the "necessary negro"', it suggests 'not only tokenism and subjugation but also acquiescence and subservience' (Powers 2012: 3). More explicitly, the reference is, using language taken up by Malcolm X, to the 'House Negro', given enough in the way of domestic comforts to identify with their enslaver.

2   An exception here is Demon Music Group, which, as the recordings and rights administration subsidiary of BBC Studios, acknowledged employees' right to become members of BECTU in alignment with other BBC employees.

Bibliography

Abbott, Andrew (1988), *The System of Professions: An Essay on the Division of Expert Labor*, Chicago: University of Chicago Press.

Adkins, Lisa (1999), 'Community and Economy: A Retraditionalization of Gender?', *Theory, Culture & Society*, 16 (1): 119–39.

Adkins, Lisa (2013), 'Creativity, Biography and the Time of Individualization', in Mark Banks, Rosalind Gill, and Stephanie Taylor (eds.), *Theorizing Cultural Work: Labour, Continuity and Change in the Cultural and Creative Industries*, 149–60, London: Routledge.

Adorno, Theodor, and Max Horkheimer (1997), *Dialectic of Enlightenment*, London: Verso.

Ahmed, Sara (2004), *The Cultural Politics of Emotion*, Edinburgh: Edinburgh University Press.

Akrich, Madeleine (1992), 'The De-scription of Technical Objects', in Wiebe E. Bijker and John Law (eds.), *Shaping Technology/Building Society*, 205–24, Cambridge: MIT Press.

Alvesson, Mats, and Hugh Wilmott (2002), 'Identity Regulation as Organizational Control: Producing the Appropriate Individual', *Journal of Management Studies*, 39 (5): 619–44.

Anand, N. (2006), 'Charting the Music Business: *Billboard* Magazine and the Development of the Commercial Music Field', in Joseph Lampel, Jamal Shamsie, and Theresa Lant (eds.), *The Business of Culture: Strategic Perspectives on Entertainment and Media*, 139–54, Mahwah: Lawrence Erlbaum.

Anderson, Benedict (2006), *Imagined Communities: Reflections on the Origins and Spread of Nationalism*, London: Verso.

Anderson, Chris (2008a), 'The End of Theory: The Data Deluge Makes the Scientific Method Obsolete', *Wired*, 23 June.

Anderson, Chris (2008b), *The Long Tail: Why the Future of Business is Selling Less of More*, 2nd edn, New York: Hyperion.

Arditi, David (2020), *Getting Signed: Record Contracts, Musicians, and Power in Society* (London: Palgrave)

Arditi, David (2021), 'The Global Music Report: Selling a Narrative of Decline', in Richard Osborne and Dave Laing (eds.), *Music By Numbers: The Use and Abuse of Statistics in the Music Industries*, 74–89, Bristol: Intellect Books.

Ariztia, Tomas (2013), 'Unpacking Insight: How Consumers are Qualified by Advertising Agencies', *Journal of Consumer Culture*, 15 (2): 143–62.

Ashton, Daniel, and Caitriona Noonan (2013), *Cultural Work and Higher Education*, London: Palgrave.

Ashton, Daniel, and Seth Giddings (2018), 'At Work in the Toybox: Bedrooms, Playgrounds and Ideas of Play in Creative Cultural Work', *The International Journal of Entrepreneurship and Innovation*, 19 (2): 81–9.

Atkinson, Rowland, Simon Parker, and Roger Burrows (2017), 'Elite Formation, Power and Space in Contemporary London', *Theory, Culture & Society*, 34 (5–6): 179–200.

Backhouse, Roger, and Béatrice Cherrier (2017), 'The Age of the Applied Economist: The Transformation of Economics Since the 1970s', *History of Political Economy*, 49: 1–33.

Bakker, Gerben (2011), 'Adopting the Rights-based Model: Music Multinationals and Local Music Industries Since 1945', *Popular Music History*, 6 (3): 307–43.

Bangs, Lester (1987), *Psychotic Reactions and Carburetor Dung*, New York: Alfred A Knopf.

Banks, Mark (2006), 'Moral Economy and Cultural Work', *Sociology*, 40 (3): 455–72.

Banks, Mark (2007), *The Politics of Cultural Work*, Basingstoke: Palgrave Macmillan.

Banks, Mark (2010), 'Craft Labour and Creative Industries', *International Journal of Cultural Policy*, 16 (3): 305–21.

Banks, Mark (2012), 'MacIntyre, Bourdieu and the Practice of Jazz', *Popular Music*, 31: 69–86.

Banks, Mark (2014), '"Being in the Zone" of Cultural Work', *Culture Unbound: Journal of Current Cultural Research*, 6: 241–62.

Banks, Mark (2019), 'Precarity, Biography and Event: Work and Time in the Cultural Industries', *Sociological Research Online*, 24 (4): 541–56.

Banks, Mark, and Katie Milestone (2011), 'Individualization, Gender and Cultural Work', *Gender, Work & Organization*, 18 (1): 73–89.

Banks, Mark, and Justin O'Connor (2017), 'Inside the Whale (and How to Get Out of There): Moving on from Two Decades of Creative Industries Research', *European Journal of Cultural Studies*, 20 (6): 637–54.

Banks, Mark, Rosalind Gill, and Stephanie Taylor, eds. (2013), *Theorizing Cultural Work: Labour, Continuity and Change in the Cultural and Creative Industries*, London: Routledge.

Barbarians at the Gate (1993), dir. Glenn Jordan, USA: Home Box Office.

Barbrook, Richard, and Andy Cameron (1996), 'Californian Ideology', *Science as Culture*, 6 (1): 44–72.

Barfe, Louis (2006), *Where Have All The Good Times Gone? The Rise and Fall of the Record Industry*, London: Atlantic.

Bates, Eliot (2020), 'Vinyl as Event: Record Store Day and the Value-vibrant Matter Nexus', *Journal of Cultural Economy*, 13 (6): 690–708.

Baym, Nancy (2010), *Personal Connections in the Digital Age*, Cambridge: Polity.

Baym, Nancy, Lana Swartz, and Andrea Alarcon (2019), 'Convening Technologies: Blockchain and the Music Industry', *International Journal of Communication*, 13: 402–21.

Beck, Ulrich, Anthony Giddens, and Scott Lash (1994), *Reflexive Modernization: Politics, Tradition and Aesthetics in the Modern Social Order*, Stanford: Stanford University Press.

Becker, Howard S. (1976), 'Art Worlds and Social Types', *American Behavioral Scientist*, 19 (6): 703–18.

Becker, Howard S. (1982), *Art Worlds*, Berkley: University of California Press.

Beer, David (2013), 'Genre, Boundary Drawing and the Classificatory Imagination', *Cultural Sociology*, 7 (2): 145–60.

Benner, Mary, and Joel Waldfogel (2016), 'The Song Remains the Same? Technological Change and Positioning in the Recorded Music Industry', *Strategy Science*, 1 (3): 129–47.

Bennett, Andy (2018), 'Conceptualising the Relationship Between Youth, Music and DIY Careers: A Critical Overview', *Cultural Sociology*, 12 (2): 140–55.

Bennett, Toby (2015), *Learning the Music Business: Evaluating the "Vocational" Turn in Music Industry Education*, London: UK Music.

Bennett, Toby (2018a), '"Essential – Passion for Music": Affirming, Critiquing and Practising Passionate Work in Creative Industries', in Lee Martin and Nick Wilson (eds.), *The Palgrave Handbook of Creativity at Work*, 431–59, London: Palgrave.

Bennett, Toby (2018b), '"The Whole Feminist-Taking-Your-Clothes-Off Thing": Negotiating the Critique of Gender Inequality in UK Music Industries', *IASPM@ Journal*, 8 (1): 24–41.

Bennett, Toby (2020a), 'Towards "Embedded Non-creative Work": Administration, Digitisation and the Recorded Music Industry', *International Journal of Cultural Policy*, 26 (2): 223–38.

Bennett, Toby (2020b), 'The Justification of a Music City: Handbooks, Intermediaries and Value Disputes in a Global Policy Assemblage', *City, Culture & Society*, 22: 21–9.

Bennett, Tony, Simon Frith, Lawrence Grossberg, John Shepherd, and Graeme Turner, eds. (1993), *Rock and Popular Music: Politics, Policies, Institutions*, London and New York: Routledge.

Beynon, Huw (1993), *Working for Ford*, London: Penguin.

Bilton, Chris (2015), 'Uncreativity: The Shadow Side of Creativity', *International Journal of Cultural Policy*, 21 (2): 153–67.

Birch, Kean, and D. T. Cochrane (2022), 'Big Tech: Four Emerging Forms of Digital Rentiership', *Science as Culture*, 31 (1): 44–58.

Birchall, Claire (2006), *Knowledge Goes Pop: From Conspiracy Theory to Gossip*, Oxford: Berg.

Böhm, Steffen, and Chris Land (2012), 'The New "Hidden Abode": Reflections on Value and Labour in the New Economy', *The Sociological Review*, 60 (2): 217–40.

Boltanski, Luc, and Ève Chiapello (2005), *The New Spirit of Capitalism*, London: Verso.

Bonini, Tiziano, and Alessandro Gandini (2019), '"First Week Is Editorial, Second Week Is Algorithmic": Platform Gatekeepers and the Platformization of Music Curation', *Social Media + Society*, 5 (4): 1–11.

Born, Georgina (1995), *Rationalizing Culture: IRCAM, Boulez and the Institutionalization of the Avant-Garde*, Berkeley: University of California Press.

Born, Georgina (2004), *Uncertain Vision: Birt, Dyke, and the Reinvention of the BBC*, London: Vintage.

Born, Georgina (2010), 'The Social and the Aesthetic: For a Post-Bourdieuian Theory of Cultural Production', *Cultural Sociology*, 4 (2): 171–208.

Born, Georgina, ed. (2013), *Music, Sound and Space: Transformations of Public and Private Experience*, Cambridge: Cambridge University Press.

Born, Georgina (2015), 'Making Time: Temporality, History and the Cultural Object', *New Literary History*, 46 (3): 361–86.

Born, Georgina, and Peter Szczepanik (2013), 'On the Ethnography of Media Production: An Interview with Georgina Born', *Iluminace*, 25 (3): 99–119.

Born, Georgina, Igor Contreras, and Annelies Fryberger (2016), '"My Responsibility Is to be Bold": An Interview with Georgina Born', *Transposition: Musique et Sciences Sociales*, 6.

Bottomley, Andrew (2015), '"Home Taping Is Killing Music": The Recording Industries' 1980s Anti-home Taping Campaigns and Struggles over Production, Labor and Creativity', *Creative Industries Journal*, 8 (2): 123–45.

Bourdieu, Pierre (1984), *Distinction: A Social Critique of the Judgment of Taste*, Cambridge: Harvard University Press.

Bourdieu, Pierre (1993), *The Field of Cultural Production*, Cambridge: Polity.

Bowen, David E., and Caren Siehl (1992), 'Sweet Music: Grateful Employees, Grateful Customers, "Grate" Profits', *Journal of Management Inquiry*, 1 (2): 154–6.

BPI (2013a), *Digital Music Nation: The UK's Digital Music Landscape*, London: BPI.

BPI (2013b), 'Record Labels Boost Diversity with Internship Scheme for Next Gen Execs'. Available online: http://www.bpi.co.uk/home/record-labels-boost-diversity -with-internship-scheme-for-next-gen-execs.aspx.

Bradshaw, Alan, and Linda M. Scott (2018), *Advertising Revolution: The Story of a Song, from Beatles Hit to Nike Slogan*, London: Repeater.

Braun, Benjamin (2021), 'Asset Manager Capitalism as a Corporate Governance Regime', in Jacob Hacker, Alexander Hertel-Fernandez, Paul Pierson, and Kathleen Thelen (eds.), *The American Political Economy: Politics, Markets, and Power*, 270–94, Cambridge: Cambridge University Press.

Braverman, Harry (1998), *Labor and Monopoly Capital: The Degradation of Work in the Twentieth Century*, New York: Monthly Review Press.

Brennan, Matt (2017), *When Genres Collide: Down Beat, Rolling Stone and the Struggle between Jazz and Rock*, London: Bloomsbury.

Burawoy, Michael (1979), *Manufacturing Consent*, Chicago: Chicago University Press.

Burkart, Patrick (2010), *Music and Cyberliberties*, Middleton: Wesleyan University Press.

Burkart, Patrick and Tom McCourt (2004), 'Infrastructure for the Celestial Jukebox', *Popular Music*, 23 (3): 349–362.

Caldwell, John T. (2003), 'Industrial Geography Lessons: Socio-professional Rituals and the Borderlands of Production Culture', in Nick Couldry and Anna McCarthy (eds.), *MediaSpace: Place, Space and Culture in a Media Age*, 163–89, London: Routledge.

Caldwell, John T. (2008), *Production Culture: Industrial Reflexivity and Critical Practice in Film and Television*, Durham: Duke University Press.

Çalişkan, Koray, and Michel Callon (2010), 'Economization, Part 2: A Research Programme for the Study of Markets', *Economy and Society*, 39 (1): 1–32.

Cammaerts, Bart, and Bingchun Meng (2011), *Creative Destruction and Copyright Protection: Regulatory Responses to File-sharing*, London: LSE.

Carah, Nicholas (2014), 'Brand Value: How Affective Labour Helps Create Brands', *Consumption Markets & Culture*, 17 (4): 346–66.

Carpentier, Nico (2015), 'Differentiating Between Access, Interaction and Participation', *Conjuctions: Transdisciplinary Journal of Cultural Participation*, 2: 9–28.

Caves, Richard (2000), *Creative Industries: Contracts Between Art and Commerce*, Cambridge: Harvard University Press.

CBI (2010), *Creating Growth: A Blueprint for the Creative Industries*, London: Confederation of British Industry.

CC Skills (2011), *The Music Blueprint*. Available online: http://www.ccskills.org.uk/Home/tabid/36/Default.aspx.

Charnas, Dan (2011), *The Big Payback: The History of the Business of Hip-Hop*, London: Penguin.

Chiapello, Ève, and Norman Fairclough (2002), 'Understanding the New Management Ideology: A Transdisciplinary Contribution from Critical Discourse Analysis and New Sociology of Capitalism', *Discourse & Society*, 13: 185–208.

Christensen, Clayton (1997), *The Innovator's Dilemma: When New Technologies Cause Great Firms to Fail*, Boston: Harvard Business School Press.

Clack, Justin (2012), *Kensington Record Labels*, London: Frost Meadowcroft.

Cloonan, Martin (2005), 'What Is Popular Music Studies? Some Observations', *British Journal of Music Education*, 22 (1): 77–93.

Cloonan, Martin (2007), *Popular Music and the State in the UK: Culture, Trade or Industry?* Aldershot: Ashgate.

Cloonan, Martin (2012), 'Selling the Experience: The World-views of Concert-promoters', *Creative Industries Journal*, 5 (1–2): 151–70.

Cloonan, Martin, and Lauren Hulstedt (2013), 'Looking for Something New: The Provision of Popular Music Studies Degrees in the UK', *IASPM@Journal*, 3 (2): 63–77.

Cluley, Robert (2011), *Creative Production in the UK Music Industries*, Unpublished PhD thesis, Leicester: University of Leicester.

CMU (n.d.) 'DCMS Economics of Streaming Inquiry', *CMU Library*. Available online: https://cmulibrary.com/timeline-dcmsstreaminginquiry/.

CMU Editorial (2014), 'Blurred Lines: Does Pop Have a Misogyny Problem?', *Complete Music Update*, 16 May. Available online: http://www.completemusicupdate.com/article/blurred-lines-does-pop-have-a-misogyny-problem/.

Cohen, Rich (2004), *Record Men: The Chess Brothers and the Birth of Rock and Roll*, New York: Norton.

Cohen, Nicole S., and Greig de Peuter (2020), *New Media Unions: Organizing Digital Journalists*, London: Routledge.

Colbourne, Rick (2011), *Practice, Power and Learning in UK Recorded Music Companies*, Unpublished PhD thesis, Cambridge: University of Cambridge.

Collini, Stefan (2012), *What Are Universities For?* London: Penguin.

Collins, Randal (2019), *The Credential Society: An Historical Sociology of Education and Stratification*, 2nd edn, New York: Columbia University Press.

Comunian, Roberta, Calvin Taylor, and David N. Smith (2014), 'The Role of Universities in the Regional Creative Economies of the UK: Hidden Protagonists and the Challenge of Knowledge Transfer', *European Planning Studies*, 22 (12): 2456–76.

Connor, Steven (2014), 'Sadistic Listening', paper delivered to John E. Sawyer Seminar, *Hearing Modernity*, Harvard University, 14 April. Available online: https://stevenconnor.com/sadisticlistening.html.

Connor, J. D. (2015), 'The Sony Hack: Data and Decision in the Contemporary Studio', *Media Industries*, 2: 2.

Consider Solutions (2016), 'The SAP Finance Journey at a Global Entertainment Group', *YouTube* Posted 12 January. Available online: https://youtu.be/hMBgFyvY8QI.

Cooke, Chris (2013), 'Record Industry Hits Out after Arts Council Chief Says New Fund Is Required to Counter Market Failure', *CMU*. Available online: http://www.completemusicupdate.com/article/record-industry-hits-out-after-arts-council-chief-says-new-fund-is-required-to-counter-market-failure/.

Cooren, Francois (2010), *Action and Agency in Dialogue: Passion, Incarnation and Ventriloquism*, Philadelphia: John Benjamins Publishing.

Costas, Jana, and Dan Kärreman (2016), 'The Bored Self in Knowledge Work', *Human Relations*, 69 (1): 61–83.

Cottom, Tressie McMillan (2017), *Lower Ed: The Troubling Rise of For-profit Colleges in the New Economy*, New York: The New Press.

Creative Society (2015), *Fair Access Principle*, London: The Creative Society.

Cronin, Anne (2004), 'Currencies of Commercial Exchange: Advertising Agencies and the Promotional Imperative', *Journal of Consumer Culture*, 4 (3): 339–60.

Cronin, Anne (2006), 'Advertising and the Metabolism of the City: Urban Space, Commodity Rhythms', *Environment and Planning D: Society and Space*, 24: 615–22.

Curran, James (2010), 'Technology Foretold', in Natalie Fenton (ed.), *New Media, Old News: Journalism and Democracy in the Digital Age*, 19–34, London: Sage.

Dannen, Frederick (1990), *Hit Men: Power Brokers and Fast Money Inside the Music Business*, New York: Random House.

David, Matthew (2010), *Peer to Peer and the Music Industry: The Criminalization of Sharing*, London: Sage.

Davies, Karen (2001), 'Responsibility and Daily Life: Reflections Over Timespace', in Jon May and Nigel Thrift (eds.), *Timespace: Geographies of Temporality*, 133–48, London: Routledge.

Davies, William (2017), *The Limits of Neoliberalism: Authority, Sovereignty and the Logic of Competition*, London: Sage.

Davis, Aeron (2017), 'Sustaining Corporate Class Consciousness Across the New Liquid Managerial Elite in Britain', *The British Journal of Sociology*, 68 (2): 234–53.

DCMS (1998), *Creative Industries Mapping Document*, London: HMSO.

DCMS (2008), *Creative Britain*, London: Department for Culture, Media and Sport.

DCMS (2014), *Creative Industries Economic Estimates January 2014*, London: HMSO.

DCMS (2015), *Creative Industries: Focus on Employment*, London: DCMS.

Deleuze, Gilles (1992), 'Postscript on the Societies of Control', *October*, 59: 3–7.

Deleuze, Gilles, and Félix Guattari (2004), *A Thousand Plateaus: Capitalism and Schizophrenia, Volume 2*, London: Continuum.

DeNora, Tia (2004), *Music in Everyday Life*, Cambridge: Cambridge University Press.

Dent Coad, Emma (2017), *After Grenfell. Housing and Inequality in Kensington and Chelsea: "The Most Unequal Borough in Britain"*, London: Royal Borough of Kensington & Chelsea.

De Peuter, Greig (2014), 'Beyond the Model Worker: Surveying a Creative Precariat', *Culture Unbound: Journal of Current Cultural Research*, 6: 263–84.

De Peuter, Greig, Nicole S. Cohen, and Enda Brophy (2015), 'Interrogating Internships: Unpaid Work, Creative Industries and Higher Education', *TripleC*, 13 (2): 329–35.

Derrida, Jacques (1994), *Specters of Marx: The State of the Debt, the Work of Mourning and the New International*, London: Routledge.

Devine, Kyle, and Alexandrine Boudreault-Fournier (2021), *Audible Infrastructures*, Oxford: Oxford University Press.

Doggett, Peter (2007), *There's a Riot Going On: Revolutionaries, Rock Stars, and the Rise and Fall of '60s Counter-Culture*, Edinburgh: Canongate.

Doggett, Peter (2009), *You Never Give Me Your Money: The Battle for the Soul of the Beatles*, London: The Bodley Head.

Du Gay, Paul (1996), *Consumption and Identity at Work*, London: Sage.

Du Gay, Paul (1998), *Production of Culture/Cultures of Production*, London: Sage.

Du Gay, Paul (2004), '"Against Enterprise" (but Not against "Enterprise", for That Would Make No Sense)', *Organization*, 11 (1): 37–57.

Du Gay, Paul (2007), *Organizing Identity: Persons and Organizations "After Theory"*, London: Sage.

Du Gay, Paul (2017), 'Office as a Vocation', *European Journal of Cultural and Political Sociology*, 4 (2): 156–65.

Du Gay, Paul, and Keith Negus (1994), 'The Changing Sites of Sound: Music Retailing and the Composition of Consumers', *Media, Culture & Society*, 16: 395–413.

Du Gay, Paul, Stuart Hall, Linda Janes, Hugh Mackay, and Keith Negus (1997), *Doing Cultural Studies: The Story of the Sony Walkman*, London: Sage.

Dumbreck, Allan, Kwela Sabine Hermanns, and Kate McBain (2003), *Sounding Out the Future: Key Skills, Training, and Education in the Music Industry*, Paisley: National Music Council.

Edwards, Lee, and Magda Pieczka (2013), 'Public Relations and "Its" Media: Exploring the Role of Trade Media in the Enactment of Public Relations' Professional Project', *Public Relations Inquiry*, 2 (1): 5–25.

Edwards, Lee, Bethany Klein, David Lee, Giles Moss, and Fiona Philip (2015), 'Discourse, Justification and Critique: Towards a Legitimate Digital Copyright Regime?', *International Journal of Cultural Policy*, 21: 60–77.

Edwards-Jones, Imogen (2009), *Pop Babylon*, London: Corgi.

Edrisi, Sam (2020), '"Where the Magic People Gathered": The Role of Private Members' Clubs in the Contemporary Music Economy', *International Journal of Music Business Research*, 9 (1): 81–117.

Eikhof, Doris Ruth and Chris Warhurst (2013), 'The Promised Land? Why Social Inequalities are Systemic in the Creative Industries', *Employee Relations*, 35 (5): 495–508.

Elberse, Anita (2010), 'Bye-Bye Bundles: The Unbundling of Music in Digital Channels', *Journal of Marketing*, 74: 107–23.

Eliot, Marc (1996), *Rockonomics: The Money Behind The Music*, New York: Citadel.

Elton, Serona (2008), 'Identifiers Used in the Music Industry', *Journal of the Music & Entertainment Industry Educators Association*, 8 (1). Available online: http://www .meiea.org/resources/Journal/html_ver/Vol08_No01/Elton-2008-MEIEA-Journal -Vol-8-No-1-p49.htm.

Emap (2003), 'Project Phoenix 2', *Emap Advertising*. Archived by Wayback Machine, captured on 10 November, 2006. Available online: https://web.archive.org/web /20061110175200/http:/www.emapadvertising.com/insight/project_detail.asp ?TypeID=2&CaseStudyID=163.

Entwistle, Joanne, and Elizabeth Wissinger (2006), 'Keeping Up Appearances: Aesthetic Labour in the Fashion Modelling Industries of London and New York', *The Sociological Review*, 54 (4): 774–94.

Eriksson, Maria, Rasmus Fleischer, Anna Johansson, Pelle Snickars, and Patrick Vonderau (2019), *Spotify Teardown: Inside the Black Box of Streaming Music*, Cambridge: MIT Press.

Etzkowitz, Henry, and Loet Leydersdorff, eds. (1997), *Universities and the Global Knowledge Economy: A Triple-Helix of University-Industry-Government Relations*, London: Pinter.

fanBase (n.d.), 'Sony Music', Archived by Wayback Machine, captured on 28 April, 2016. Available online: https://web.archive.org/web/20160428233430/http://fanbase .sonymusic.co.uk:80/.

Farocki, Harun (2001), 'Workers Leaving The Factory', trans. Laurent Faasch-Ibrahim, *Senses of Cinema*, 21 February. Available online: https://yaleunion.org/wp-content/ uploads/2013/12/farocki_workersleaving.pdf.

Figiel, Joanna, Stevphen Shukaitis, and Abe Walker (2014), 'The Politics of Workers' Inquiry', *ephemera*, 14 (3): 307–14.

Finnegan, Ruth (1989), *The Hidden Musicians: Music-Making in an English Town*, Cambridge: Cambridge University Press.

Fisher, Mark (2014), *Ghosts of My Life: Writings on Depression, Hauntology and Lost Futures*, London: Repeater.

Fiske, John (1992), 'The Cultural Economy of Fandom', in Lisa A. Lewis (ed.), *The Adoring Audience: Fan Culture and Popular Media*, 30–49, London: Routledge.

Fleming, Peter (2015), *The Mythology of Work: How Capitalism Persists Despite Itself*, London: Pluto Press.

Florida, Richard (2002), *Rise of the Creative Class*, New York: Basic Books.

Forde, Eamonn (2000), *Music Journalists, Music Press Officers & the Consumer Press in the UK*, Unpublished PhD thesis, University of Westminster.

Forde, Eamonn (2001), 'From Polyglottism to Branding: On the Decline of Personality Journalism in the British Music Press', *Journalism*, 2 (1): 23–43.

Forde, Eamonn (2006), 'Conflict and Collaboration: The Press Officer/Journalist Nexus in the British Music Press of the Late 1990s', *Popular Music History*, 1 (3): 285–306.

Forde, Eamonn (2015), 'The 15 Music Industry Books You Must Read', *Music Ally*. Available online: http://musically.com/2015/07/29/15-music-industry-books-must-read/.

Forde, Eamonn (2019), *The Final Days of EMI: Selling the Pig*, London: Omnibus.

Frank, Thomas (1997), *The Conquest of Cool: Business Culture, Counterculture, and the Rise of Hip Consumerism*, Chicago: University of Chicago Press.

Fraser, Nancy (2013), *Fortunes of Feminism: From State-managed Capitalism to Neoliberal Crisis*, London: Verso.

Frenette, Alexandre (2013), 'Making the Intern Economy: Role and Career Challenges of the Music Industry Intern', *Work and Occupations*, 40: 364–97.

Frith, Simon (1981), *Sound Effects: Youth, Leisure and the Politics of Rock 'n' Roll*, New York: Pantheon.

Frith, Simon (1986), 'Art Versus Technology: The Strange Case of Popular Music', *Media, Culture & Society*, 8: 263–79.

Frith, Simon (1996), *Performing Rites: On the Value of Popular Music*, Cambridge: Harvard University Press.

Frith, Simon (2000), 'Music Industry Research: Where Now? Where Next? Notes from Britain', *Popular Music*, 19 (3): 387–93.

Frith, Simon (2007), 'Live Music Matters', *Scottish Music Review*, 1 (1): 1–17.

Frith, Simon (2017), 'Are Workers Musicians?', *Popular Music*, 36 (1): 111–15.

Frith, Simon (2019), 'Remembrance of Things Past: Marxism and the Study of Popular Music', *Twentieth-Century Music*, 16 (1): 141–55.

Frith, Simon, and Andrew Goodwin (1990), *On Record: Rock, Pop and the Written Word*, London: Routledge.

Frith, Simon, and Howard Horne (1987), *Art into Pop*, London: Routledge.

Fukuyama, Francis (1992), *The End of History and The Last Man*, New York: The Free Press.

Fuller, Matthew (2005), *Media Ecologies: Materialist Energies in Art and Technoculture*, Cambridge: MIT Press.

Galloway, Alexander R. (2013), 'The Poverty of Philosophy: Realism and Post-fordism', *Critical Inquiry*, 39: 347–66.

Gieryn, Thomas (1983), 'Boundary-Work and the Demarcation of Science from Non-Science: Strains and Interests in Professional Ideologies of Scientists', *American Sociological Review*, 48 (6): 781–95.

Gil, Natalie (2014), 'Simon Cowell's Company Criticised for Advertising for Unpaid Interns', *The Guardian*. Available online: https://www.theguardian.com/media/2014/jan/22/simon-cowell-company-criticised-advertising-unpaid-interns.

Gilbert, Jeremy (2012), 'Capitalism, Creativity and the Crisis in the Music Industry', *Open Democracy UK*. Available online: http://www.opendemocracy.net/ourkingdom/jeremy-gilbert/capitalism-creativity-and-crisis-in-music-industry.

Gilbert, Jeremy, and Ewan Pearson (1999), *Discographies: Dance Music, Culture and the Politics of Sound*, London: Routledge.

Gildart, Keith (2013), '"The Antithesis of Humankind": Exploring Responses to the Sex Pistols' Anarchy Tour 1976', *Cultural and Social History*, 10 (1): 129–429.

Gill, Rosalind (2002), 'Cool, Creative and Egalitarian? Exploring Gender in Project-based New Media Work in Europe', *Information, Communication & Society*, 5 (1): 70–89.

Gill, Rosalind (2003), 'Power and the Production of Subjects: A Genealogy of the New Man and the New Lad', *The Sociological Review*, 51: 34–56.

Gill, Rosalind (2009), 'Creative Biographies in New Media: Social Innovation in Web Work', in Andy Pratt and Paul Jeffcutt (eds.), *Creativity, Innovation and the Cultural Economy*, 161–78, Oxford: Routledge.

Gill, Rosalind (2014), 'Unspeakable Inequalities: Post Feminism, Enterpreneurial Subjectivity, and the Repudiation of Sexism Among Cultural Workers', *Social Politics*, 21 (4): 509–28.

Gill, Rosalind, and Andy Pratt (2008), 'In the Social Factory? Immaterial Labour, Precariousness and Cultural Work', *Theory, Culture and Society*, 25 (7–8): 1–30.

Goodell Brown, Robert (1959), *Statistical Forecasting for Inventory Control*, New York: McGraw Hill.

Goodman, Fred (2010), *Fortune's Fool: Edgar Bronfman Jr., Warner Music, and an Industry in Crisis*, New York: Simon & Schuster.

Graeber, David (2013), *Bullshit Jobs: A Theory*, London: Penguin.

Grainge, Lucian (2009), 'Digital Music Talent Deserves Reward', *The Guardian*. Available online: https://www.theguardian.com/business/2009/aug/29/universal -music-downloads-digital.

Gray, Robert (2007), 'Market Research: Reaching Out to the Kids', *Campaign*, 28 November.

Greenfield, Robert (2011), *The Last Sultan: The Life and Times of Ahmet Ertegun*, London: Simon & Schuster.

Gregg, Melissa (2010), 'On Friday Night Drinks: Workplace Affects in the Age of the Cubicle', in Melissa Gregg and Gregory J. Seigworth (eds.), *The Affect Theory Reader*, 250–68, Durham: Duke University Press.

Gregg, Melissa (2018), *Counterproductive: Time Management in the Knowledge Economy*, Durham: Duke University Press.

Gross, Jonathan (2020), *The Birth of the Creative Industries Revisited: An Oral History of the 1998 DCMS Mapping Document*, London: KCL.

Gross, Sally Anne, and George Musgrave (2020), *Can Music Make You Sick? Measuring the Price of Musical Ambition*, London: University of Westminster Press.

Guichardaz, Rémy, Laurent Bach, and Julien Penin (2019), 'Music Industry Intermediation in the Digital Era and the Resilience of the Majors' Oligopoly: The Role of Transactional Capability', *Industry and Innovation*, 26 (7): 843–69.

Habermas, Jurgen (1975), *Legitimation Crisis*, Cambridge: Polity Press.

Hagen, Anya (2015), 'The Playlist Experience: Personal Playlists in Music Streaming Services', *Popular Music and Society*, 38 (5): 625–45.

Hall, Stuart and Paddy Whannel (1964), *The Popular Arts*, London: Hutchinson.

Harker, Dave (1980), *One for the Money: Politics and Popular Song*, London: Hutchison.

Harker, Dave (1997), 'The Wonderful World of IFPI: Music Industry Rhetoric, the Critics and the Classical Marxist Technique', *Popular Music*, 16 (1): 45–79.

Harrington, Brooke (2003), 'The Social Psychology of Access in Ethnographic Research', *Journal of Contemporary Ethnography*, 32 (5): 592–625.

Harris, John (2009), 'The Pirates Thrive on a Scrap with the Analogue Crowd', *The Guardian*. Available online: https://www.theguardian.com/commentisfree/2009/aug /26/john-harris-piracy-business-pragmatism.

Harris, Thomas A. (1969), *I'm Ok, You're OK*, London: Harper & Row.

Harrison, Ann (2014), *Music: The Business. The Essential Guide to the Law and the Deals*, 6th edn, London: Virgin.

Haynes, Jo, and Lee Marshall (2018), 'Reluctant Entrepreneurs: Musicians and Entrepreneurship in the "New" Music Industry', *The British Journal of Sociology*, 69 (2): 459–82.

Hebdige, Dick (1979), *Subculture: The Meaning of Style*, London: Methuen.

Heelas, Paul (2002), 'Work Ethics, Soft Capitalism and the "Turn to Life"', in Paul du Gay and Michael Pryke (eds.), *Cultural Economy: Cultural Analysis and Commercial Life*, , 78–96, London: Sage.

Heller, Michael (2008), *Gridlock Economy: How Too Much Ownership Wrecks Markets, Stops Innovation and Costs Lives*, New York: Basic Books.

Hennion, Antoine (1989), 'An Intermediary Between Production and Consumption: The Producer of Popular Music', *Science, Technology & Human Values*, 14 (4): 400–24.

Hennion, Antoine (2007), 'Those Things that Hold us Together: Taste and Sociology', *Cultural Sociology*, 1: 97–114.

Hennion, Antoine (2010), 'Loving Music: From a Sociology of Mediation to a Pragmatics of Taste', *Comunicar*, 34 (17): 25–33.

Hermida, Alfred (2010), 'Twittering the News: The Emergence of Ambient Journalism', *Journalism Practice* 4 (3): 297–308.

Hesmondhalgh, David (1996), 'Flexibility, Post-fordism and the Music Industries', *Media, Culture & Society*, 18: 469–88.

Hesmondhalgh, David (1998a), 'Post-Punk's Attempt to Democratise the Music Industry: The Success and Failure of Rough Trade', *Popular Music*, 16 (3): 255–74.

Hesmondhalgh, David (1998b), 'The British Dance Music Industry: A Case Study of Independent Cultural Production', *The British Journal of Sociology*, 49 (2): 234–51.

Hesmondhalgh, David (2005), 'Subcultures, Scenes or Tribes? None of the Above', *Journal of Youth Studies*, 8 (1): 21–40.

Hesmondhalgh, David (2007), *The Cultural Industries*, London: Sage.

Hesmondhalgh, David (2009), 'The Digitalisation of Music', in Andy C. Pratt and Paul Jeffcutt (eds.), *Creativity, Innovation and the Cultural Economy*, 57–73, Oxford: Routledge.

Hesmondhalgh, David (2013a), *Why Music Matters*, London: Wiley-Blackwell.

Hesmondhalgh, David (2013b), *The Cultural Industries*, 3rd edn, London: Sage.

Hesmondhalgh, David (2014), 'The Menace of Instrumentalism in Media Industries Research and Education', *Media Industries Journal*, 1 (1): 21–6.

Hesmondhalgh, David (2021), 'Is Music Streaming Bad for Musicians? Problems of Evidence and Argument', *New Media & Society*, 23 (12): 3593–615.

Hesmondhalgh, David (2022), 'Streaming's Effects on Music Culture: Old Anxieties and New Simplifications', *Cultural Sociology*, 16 (1): 3–24.

Hesmondhalgh, David, and Leslie M. Meier (2018), 'What the Digitalisation of Music Tells Us About Capitalism, Culture and the Power of the Information Technology Sector', *Information, Communication & Society*, 21 (11): 1555–70.

Hill, Keith (1999), 'A Perspective: The Role of Identifiers in Managing and Protecting Intellectual Property in the Digital Age', *Proceedings of the IEEE*, 87 (7): 1228–38.

Hirsch, Paul (1972), 'Processing Fads and Fashions: An Organization-set Analysis of Cultural Industry Systems', *American Journal of Sociology*, 77: 639–59.

Hirschman, A. O. (1970), *Exit, Voice and Loyalty: Responses to Decline in Firms, Organizations, and States*, Cambridge: Harvard University Press.

Hirschmann, A. O. (1997), *The Passions and the Interests: Political Arguments for Capitalism Before its Triumph*, Princeton: Princeton University Press.

Hits Daily Double (2012), 'Lucian Grainge on the Record', *Hits Daily Double*. Available online: http://hitsdailydouble.com/news&id=271576.

HMRC (2014), 'Brit Awards Record Labels Face the Music on Unpaid Interns', HM Revenue & Customs. Available online: https://www.gov.uk/government/news/brit-awards-record-labels-face-the-music-on-unpaid-interns (accessed 12 January 2018).

Holt, Jennifer, and Alisa Perren (2009), *Media Industries: History, Theory and Method*, Oxford: Wiley.

Homewood, Ben (2018), '"You Don't Need a Major Label to Have a Success": Why the Label Services Sector is Bigger than Ever', *MusicWeek*, 1 February. Available online: http://www.musicweek.com/labels/read/you-don-t-need-a-major-label-to-have-a-success-why-the-label-services-sector-is-bigger-than-ever/071315.

Housham, Jane (2009), 'Review: Mad, Gleeful Nastiness', *The Guardian*, 31 January. Available online: https://www.theguardian.com/books/2009/jan/31/kill-your-friends-review.

Hracs, Brian (2015), 'Cultural Intermediaries in the Digital Age: The Case of Independent Musicians and Managers in Toronto', *Regional Studies*, 49 (3): 461–75.

Hracs, Brian, and Johan Jansson (2020), 'Death by Streaming or Vinyl Revival? Exploring the Spatial Dynamics and Value-creating Strategies of Independent Record Shops in Stockholm', *Journal of Consumer Culture*, 20 (4): 478–97.

Hu, Cherie (2018), 'Spotify Wants to be "The R&D Department for the Entire Music Industry" – What Does that Actually Mean?', *Billboard*, 26 April. Available online: https://www.billboard.com/pro/spotify-rd-department-entire-music-industry-netflix/.

Hu, Cherie (2020), 'Tencent Isn't Actually a Chinese Company: Why Should the Global Music Business Care?', *Music Business Worldwide*, 11 March. Available online: https://www.musicbusinessworldwide.com/tencent-music-isnt-actually-a-chinese-company-why-should-the-global-music-business-care/.

Hughes, Diane, Mark Evans, Guy Morrow, and Sarah Keith (2016), *The New Music Industries: Disruption and Discovery*, London: Palgrave.

Huq, Rupa (2010), 'Labouring the Point? The Politics of Britpop in "New Britain"', in Andy Bennett and Jon Stratton (eds.), *Britpop and the English Music Tradition*, 89–102, Farnham: Ashgate.

Huygens, Marc, Charles Baden-Fuller, Frans A. J. Van Den Bosch, and Henk W. Volberda (2001), 'Co-evolution of Firm Capabilities and Industry Competition: Investigating the Music Industry, 1877–1997', *Organization Studies*, 22 (6): 971–1011.

IFPI (2009), *International Standard Recording Code Handbook*, London: IFPI.

IFPI (2013), *Digital Music Report: Engine of a Digital World*, London: IFPI.

IFPI (2016), *Global Music Report: Music Consumption Exploding Worldwide*, London: IFPI.

Ingham, Tim (2015), 'The Music Industry Would be Insane to Dump Freemium', *Music Business Worldwide*. Available online: http://www.musicbusinessworldwide.com/the -music-industry-would-be-insane-to-dump-freemium/.

Ingham, Tim (2017), 'Universal has Fewer Staff Than It Did 10 Years Ago – Despite Buying EMI', *Music Business Worldwide*, 26 April. Available online: https://www .musicbusinessworldwide.com/universal-has-fewer-staff-than-it-did-10-years-ago -despite-buying-emi/.

Ingham, Tim (2021), 'In Historic Move, Sony Is Disregarding Unrecouped Balances for Heritage Catalog Artists', *Music Business Worldwide*, 11 June. Available online: https://www.musicbusinessworldwide.com/in-historic-move-sony-music-announces -its-disregarding-unrecouped-balances-for-heritage-catalog-artists/.

Ingham, Tim (2023a), 'The 3 Major Music Companies Employed Over 27,000 People Between Them Last Year – Up By 6,660 On 2017', *Music Business Worldwide*, 28 June. Available online: https://www.musicbusinessworldwide.com/the-3-major -music-companies-employed-over-27000-people-between-them-last-year-up-by -over-7000-on-2017/.

Ingham, Tim (2023b), 'The Modern Record Industry Is Moving into its Cut to Grow Phase', *Music Business Worldwide*. Available online: https://www.musicbusinesswo rldwide.com/the-modern-record-industry-is-moving-into-its-cut-to-grow-phase/.

IPO (2019), *Music 2025. The Music Data Dilemma: Issues Facing the Music Industry in Improving Data Management*, Newport: Intellectual Property Office.

James, Robin (2015), *Resistance and Melancholy: Pop Music, Feminism and Neoliberalism*, Winchester: Zero Books.

Jenkins, Henry (2006), *Convergence Culture: Where Old and New Media Collide*, New York: New York University Press.

Jennings, David (2006), 'Groups and Behaviour Patterns Among Music Listeners', *DJ Alchemi*. Available online: http://alchemi.co.uk/archives/mus/groups_and_beha.html.

Jones, Mike (1999), 'Changing Slides – Labour's Music Industry Policy Under the Microscope', *Critical Quarterly*, 41 (1): 22–31.

Jones, Mike (2002), 'Learning to Crawl: The Rapid Rise of Music Industry Education', in Martin Talbot (ed.), *The Business of Music*, 292–311, Liverpool: Liverpool University Press.

Jones, Mike (2012), *The Music Industries: From Conception to Consumption*, Basingstoke: Palgrave Macmillan.

Jones, Michael L. (2016), 'Revisiting "Music Industry Research": What Changed? What Didn't?', in Lee Marshall and Dave Laing (eds.), *Popular Music Matters: Essays in Honour of Simon Frith*, 45–60, Abingdon: Routledge.

Jones, Steve (2000), 'Music and the Internet', *Popular Music*, 19 (2): 217–30.

Jones, Steve (2002), 'Music that Moves: Popular Music Distribution and Network Technologies', *Cultural Studies*, 16 (2): 213–32.

Karp, Jonathan (2015), 'Brokering a Rock "n" Roll International: Jewish Record Men in America and Britain', in Rebecca Kobrin and Adam Teller (eds.), *Purchasing Power: The Economics of Modern Jewish History*, 125–52, Pennsylvania: University of Pennsylvania Press.

Keen, Andrew (2007), *The Cult of the Amateur: How Today's Internet is Killing Our Culture*, New York: Doubleday.

Kelly, Kevin (2002), 'Towards a Post-napster Music Industry', *The New York Times Magazine*, 17 March, 19–21.

Kelly, Kevin (2008), '1,000 True Fans', *The Technium*. Available online: http://kk.org/thetechnium/1000-true-fans/.

Kelly, Kim (2023), 'Bandcamp Workers Form a Union: "It's Not Enough to Get Small Wins Alone"', *Rolling Stone*, 16 March. Available online: https://www.rollingstone.com/music/music-features/bandcamp-union-1234695308/.

King, Richard (2012), *How Soon Is Now? The Madmen and Mavericks who Made Independent Music 1975–2005*, London: Faber & Faber.

Klein, Bethany (2008), '"The New Radio": Music Licensing as a Response to Industry Woe', *Media, Culture & Society*, 30 (4): 463–78.

Klein, Bethany (2017), 'Selling Out: Musicians, Autonomy and Compromise in the Digital Age', *Popular Music and Society*, 40 (2): 222–38.

Kleinman, Daniel Lee, and Steven P. Vallas (2001), 'Science, Capitalism and the Rise of the "Knowledge Worker": The Changing Structure of Knowledge Production in the United States', *Theory and Society*, 30: 451–92.

Knights, David, Theo Vurdubakis, and Hugh Wilmott (2008), 'The Night of the Bug: Technology, Risk and (Dis)organization at the *Fin de Siècle*', *Management & Organizational History*, 3 (34): 289–309.

Knorr-Cetina, Karin (2010), 'The Epistemics of Information: A Consumption Model', *Journal of Consumer Culture*, 10 (2): 171–201.

Koester, Almut (2006), *Investigating Workplace Discourse*, London: Routledge.

Koppman, Sharon (2012), 'Different Like Me: Why Cultural Omnivores Get Creative Jobs', *Administrative Science Quarterly*, 61 (2): 291–331.

Korczynski, Marek, Michael Pickering, and Emma Robertson (2013), *Rhythms of Labour: Music at Work in Britain*, Cambridge: Cambridge University Press.

Kretschmer, Martin, and Andy C. Pratt (2009), 'Music, Copyright and Information and Communications Studies', *Information, Communication & Society*, 12 (2): 165–77.

Kribs, Kait (2017), 'The Artist-as-Intermediary: Musician Labour in the Digitally Networked Era', *ETopia*. https://doi.org/10.25071/1718-4657.36768.

Kusek, David, and Gerd Leonhard (2005), *The Future of Music: Manifesto for the Digital Music Revolution*, Boston: Berklee Press.

Lamere, Paul (2014), 'The Zero Button Music Player', *Music Machinery*. Available online: https://musicmachinery.com/2014/01/14/the-zero-button-music -player-2/.

Latham, Sean (2009), *The Art of Scandal: Modernism, Libel Law and the Roman à Clef*, Oxford: Oxford University Press.

Latour, Bruno (2004), 'Why Has Critique Run out of Steam? From Matters of Fact to Matters of Concern', *Critical Inquiry*, 30: 225–48.

Latour, Bruno (2005), *Reassembling the Social: An Introduction to Actor-Network Theory*, Oxford: Oxford University Press.

Law, John (1994), *Organizing Modernity*, Oxford: Blackwell.

Law, John (2002), 'Economics as Interference', in Paul du Gay and Michael Pryke (eds.), *Cultural Econony: Cultural Analysis and Commercial Life*, 39–58, London: Sage.

Lea, Tom (2015), 'Inside Album Leaks: How Do They Happen, How Do We Stop Them, and Do They Even Matter?', *FACT*. Available online: http://www.factmag.com/2015 /01/29/inside-album-leaks-bjork-vulnicura-madonna-rebel-heart/.

Leadbeater, Charles, and Kate Oakley (1999), *The Independents: Britain's New Cultural Entrepreneur*, London: Demos.

Lefebvre, Henri (2004), *Rhythmanalysis: Space, Time and Everyday Life*, London: Continuum.

Leidner, Robin (2006), 'Identity and Work', in Marek Korczynski, Randy Hodson, and Paul Edwards (eds.), *Social Theory at Work*, 424–63, Oxford: Oxford University Press.

Leonard, Marion (2016), 'Girls at Work: Gendered Identities, Sex Segregation and Employment Experiences in the Music Industries', in J. Warwick and A. Adrian (eds.), *Voicing Girlhood in Popular Music: Performance, Authority, Authenticity*, 37–55, New York: Routledge.

Lepore, Jill (2014), 'The Disruption Machine', *The New Yorker*, 23 June.

Lessig, Lawrence (2004), *Free Culture: How Big Media Uses Copyright and the Law to Lock Down Culture and Control Creativity*, New York: Penguin Press.

Leyshon, Andrew (2014), *Reformatted: Code, Networks, and the Transformation of the Music Industry*, Oxford: Oxford University Press.

Leyshon, Andrew, Nigel Thrift, Louise Crewe, Shaun French, and Pete Webb (2016), 'Leveraging Affect: Mobilizing Enthusiasm and the Co-production of the Musical Economy', in Brian Hracs, Brian Seman, and Tarek Virani (eds.), *The Production and Consumption of Music in the Digital Age*, 248–62, London: Routledge.

Lindvall, Helienne (2011), 'Behind the Music: What's Happening to "On Air, On Sale"?', *The Guardian*. Available online: https://www.theguardian.com/music/musicblog /2011/aug/05/behind-music-on-air-sale-radio.

Lindvall, Helienne (2012), '"Island Records" Darcus Beese: You Sign What Makes You Excited', *The Guardian*. Available online: https://www.theguardian.com/media/2012/ jul/15/island-records-darcus-beese.

Lindvall, Helienne (2013), 'Why Unpaid Internships Are Bad for the Music Industry', *The Guardian*. Available online: https://www.theguardian.com/media/2013/aug/15/ unpaid-interns-music-industry.

Loog-Oldham, Andrew (2001), *Stoned*, New York: Vintage.

Lorde, Audre (2018), *The Master's Tools Will Never Dismantle the Master's House*, London: Penguin.

Lovell, Nicholas (2013), *The Curve: From Freeloaders into Superfans*, London: Penguin.

McCarthy, Anna (2004), 'Geekospheres: Visual Culture and Material Culture at Work', *Journal of Visual Culture*, 3 (2): 213–21.

McCloud, Sean (2003), 'Popular Culture Fandoms, the Boundaries of Religious Studies, and the Project of the Self', *Culture and Religion*, 4 (2): 187–206.

McGee, Alan (2013), *Creation Stories: Riots, Raves and Running a Label*, London: Sidgwick & Jackson.

McGettigan, Andrew (2013), *The Great University Gamble: Money, Markets and the Future of Higher Education*, London: Pluto.

McKinlay, Alan, and Chris Smith (2009), *Creative Labour: Working in the Creative Industries*, Basingstoke: Palgrave Macmillan.

McLuhan, Marshall (1964), *Understanding Media: The Extensions of Man*, New York: McGraw Hill.

McRobbie, Angela (1998), *British Fashion Design: Rag Trade or Image Industry*, London: Routledge.

McRobbie, Angela (2016), *Be Creative: Making a Living in the New Culture Industries*, Cambridge: Polity.

Mackenzie, Jean (2017), 'Rape and Abuse: The Music Industry's Dark Side Exposed', *BBC*, 18 December. Available online: http://www.bbc.co.uk/news/entertainment-arts -42368544.

Mall, Andrew (2018), 'Concentration, Diversity, and Consequences: Privileging Independent Over Major Record Labels', *Popular Music*, 37 (3): 444–65.

Mandler, Peter (2020), *The Crisis of the Meritocracy: Britain's Transition to Mass Education Since the Second World War*, Oxford: Oxford University Press.

Manovich, Lev (1999), 'Database as Symbolic Form', *Convergence*, 5 (80): 80–99.

Marcus, George (1995), 'Ethnography in/of the World System: The Emergence of Multi-sited Ethnography', *Annual Review of Anthropology*, 24: 95–117.

Marcuse, Herbert (1991), *One-Dimensional Man: Studies in the Ideology of Advanced Industrial Society*, 2nd edn, Oxford: Routledge.

Marshall, Lee (2012), 'The Structural Functions of Stardom in the Recording Industry', *Popular Music and Society*, iFirst. http://dx.doi.org/10.1080/03007766.2012.718509.

Marshall, Lee (2013), 'The 360 Deal and the "New" Music Industry', *European Journal of Cultural Studies*, 16 (1): 77–99.

Marshall, Lee (2015), '"Let's Keep Music Special: F--- Spotify": On-demand Streaming and the Controversy Over Artist Royalties', *Creative Industries Journal*, 8 (2): 177–89.

Martin, Graham (2005), 'Narratives Great and Small: Neighbourhood Change, Place and Identity in Notting Hill', *International Journal of Urban and Regional Research*, 29 (1): 67–88.

Marx, Karl (1976), *Capital, Volume 1*, London: Penguin.

Mattern, Shannon (2018), 'Maintenance and Care', *Places*. Available online: https://placesjournal.org/article/maintenance-and-care/?cn-reloaded=1#0.

Mayer, Vicki (2011), *Below the Line: Producers and Production Studies in the New Television Economy*, Durham: Duke University Press.

Mayer, Vicki, Miranda Banks, and John T. Caldwell (2009), *Production Studies: Cultural Studies of Media Industries*, London: Routledge.

Menn, Joseph (2003), *All the Rave: The Rise and Fall of Shawn Fanning's Napster*, New York: Crown Business.

Meier, Leslie (2016), *Popular Music as Promotion: Music and Branding in the Digital Age*, Oxford: Wiley.

Miles, Andrew, and Mike Savage (2012), 'The Strange Survival Story of the English Gentleman, 1945-2010', *Cultural and Social History*, 9 (4): 595–612.

Milonakis, Dimitris, and Ben Fine (2009), *From Economics Imperialism to Freakonomics: The Shifting Boundaries Between Economics and Other Social Sciences*, London: Routledge.

Minto, Barbara (2008), *The Pyramid Principle: Logic in Writing and Thinking*, 3rd edn, Harlow: Pearson Education.

Morgan, Gareth (1986), *Images of Organization*, London: Sage.

Morris, Jeremy W. (2012), 'Making Music Behave: Metadata and the Digital Music Commodity', *New Media & Society*, 14 (5): 850–66.

Morris, Jeremy W. (2014), 'Artists as Entrepreneurs, Fans as Workers', *Popular Music and Society*, 37 (3): 273–90.

Morris, Meaghan (1989), 'Banality in Cultural Studies', *Discourse*, 10 (2): 3–29.

Morrow, Guy (2022), *Rethinking the Music Business: Music Contexts, Rights, Data and Covid-19*, London: Springer.

Mort, Frank (1996), *Cultures of Consumption: Masculinities and Social Space in Late Twentieth-Century Britain*, London: Routledge.

Morgan, George, and Julian Wood (2014), 'Creative Accommodations: The Fractured Transitions and Precarious Lives of Young Musicians', *Journal of Cultural Economy*, 7 (1): 64–78.

Morwick, Jason, Robyn Bews, Emily Klein, and Tim Lorman (2013), *Workshift: Future Proof Your Organization*, London: Palgrave.

Mosser, Jason (2012), 'What's Gonzo About Gonzo Journalism?', *Literary Journalism Studies*, 4 (1): 85–90.

Moules, Jonathan (2016), 'A Music Industry MBA Is the New Rock'n'roll', *Financial Times*, 12 December. Available online: https://www.ft.com/content/4072be00-bbe5-11e6-8b45-b8b81dd5d080.

Mueller, Andrew (2005), 'Rock of Rages', *The Guardian*, 26 February. Available online: https://www.theguardian.com/music/2005/feb/26/popandrock.music.

Mulligan, Mark (2011), 'Why the Access versus Ownership Debate Isn't Going to Resolve Itself Anytime Soon', *Music Industry Blog*. Available online: https://musicindustryblog.wordpress.com/2011/12/09/why-the-access-versus-ownership -debate-isnt-going-to-resolve-itself-anytime-soon/.

Mulligan, Mark (2014), 'The Great Music Industry Power Shift', *Music Industry Blog*. Available online: http://musicindustryblog.wordpress.com/2014/06/04/the-great -music-industry-power-shift/.

Mulligan, Mark (2015a), 'The Case for a Freemium Reset', *Music Industry Blog*. Available online: https://musicindustryblog.wordpress.com/2015/03/11/the-case-for -a-freemium-reset/.

Mulligan, Mark (2015b), *Awakening: The Music Industry in the Digital Age*, London: MIDiA Research.

Muniesa, Fabian (2014), *The Provoked Economy: Economic Reality and the Performative Turn*, Abingdon: Routledge.

Muniesa, Fabian (2017), 'The Live Act of Business and the Culture of Realization', *HAU: Journal of Ethnographic Theory*, 7 (3): 347–62.

Muniesa, Fabian, and Claes-Fredrik Helgesson (2013), 'Valuation Studies and the Spectacle of Valuation', *Valuation Studies*, 1 (2): 119–23.

Munnik, Michael B. (2016), 'When You Can't Rely on Public or Private: Using the Ethnographic Self as a Resource', in Chris Paterson, David Lee, Anamik Saha, and Anna Zoellner (eds.), *Advancing Media Production Research: Shifting Sites, Methods and Politics*, 147–60, London: Palgrave.

Murphy, Gareth (2014), *Cowboys and Indies: The Epic History of the Record Industry*, London: Serpent's Tail.

Music Ally (2020), 'Catalogue Marketing Rejuvenated', *Music Ally*, 19 February. Available online: https://musically.com/2020/02/19/catalogue-marketing -rejuvenated/.

Mutrie, Eric (2022), 'Sony Music UK's New London Office Shows Off Rich Sonic Range', *Azure*, 31 October. Available online: https://www.azuremagazine.com/article/sony -music-uk-kings-cross-london-moreysmith/.

Napier-Bell, Simon (2001), *Black Vinyl, White Powder*, London: Ebury Press.

Napier-Bell, Simon (2008), 'The Life and Crimes of the Music Biz', *The Observer*. Available online: http://www.theguardian.com/music/2008/jan/20/popandrock .musicindustry.

Napier-Bell, Simon (2015), *Ta Ra Ra Boom De-Ay: The Dodgy Business of Popular Music*, London: Unbound.

Negroponte, Nicholas (1996), *Being Digital*, New York: Vintage.

Negus, Keith (1992), *Producing Pop*, London: Edward Arnold.

Negus, Keith (1999), *Music Genres and Corporate Cultures*, London: Routledge.

Negus, Keith (2002), 'The Work of Cultural Intermediaries and the Enduring Distance Between Production and Consumption', *Cultural Studies*, 16 (4): 501–15.

Negus, Keith (2015), 'Recordings, Rights and Risks: Intermediaries and the Changing Music Industries', *Civilisations: Revue Internationale d'Anthropologie et de Sciences Humaine*, 13: 113–36.

Negus, Keith (2021), 'Perspectives on Cultural Economy: Personal, Institutional, Historical', *Journal of Cultural Economy*. https://doi.org/10.1080/17530350.2021.1958900.

Newkey-Burden, Chaz (2009), *Simon Cowell: The Unauthorized Biography*, London: Michael O'Mara.

Nicolau, Anna, and Kaye Wiggins (2022), 'How Wall Street Stormed the Music Business', *Financial Times*, 13 September. Available online: https://www.ft.com/content/e879f856-3ec3-4bd7-b564-ada5b590e3ef.

Niven, John (2008), "Til Death Did We Party: The Music Business in the Heady Nineties', *The Times*, 15 February. Available online: https://www.thetimes.co.uk/article/til-death-did-we-party-the-music-business-in-the-heady-nineties-c7p9b3tnlld.

Niven, John (2009), *Kill Your Friends*, London: Vintage.

Nixon, Sean (1996), *Hard Looks: Masculinities, Spectatorship and Contemporary Consumption*, New York: St Martin's Press.

Nixon, Sean (2003), *Advertising Cultures: Gender, Commerce, Creativity*, London: Sage.

Nixon, Sean, and Ben Crewe (2004), 'Pleasure at Work? Gender, Consumption and Work-based Identities in the Creative Industries', *Consumption Markets & Culture*, 7 (2): 129–47.

Nordgård, Daniel (2018), *The Music Business and Digital Impacts: Innovations and Disruptions in the Music Industries*, Cham: Springer Nature.

Nowak, R. (2016), *Consuming Music in the Digital Age: Technologies, Roles and Everyday Life*, London: Palgrave.

O'Connor (2022), 'Music as Industry', *Wake In Fright*, 3 April. Available online: https://wakeinalarm.blog/2021/04/03/music-as-industry/.

Oakley, Kate (2013), 'Absentee Workers: Representation and Participation in the Cultural Industries', in Mark Banks, Rosalind Gill, and Stephanie Taylor (eds.), *Theorizing Cultural Work: Labour, Continuity and Change in the Creative Industries*, 56–67, London: Routledge.

Oakley, Kate, Daniel Laurison, Dave O'Brien, and Sam Friedman (2017), 'Cultural Capital: Arts Graduates, Spatial Inequality, and London's Impact on Cultural Labor Markets', *American Behavioral Scientist*, 61 (12): 1510–31.

Oberholzer-Gee, Felix, and Koleman Strumpf (2007), 'The Effect of File Sharing on Record Sales: An Empirical Analysis', *Journal of Political Economy*, 115 (1): 1–42.

OCC (2019), 'Official Charts Announces the Noughties Definitive Chart Books', *Official Charts Company*, 24 September. Available online: https://www.officialcharts.com/chart-news/official-charts-announces-the-noughties-definitive-chart-books__27299/.

Ortner, Sherry (2010), 'Access: Reflections on Studying Up in Hollywood', *Ethnography*, 11 (2): 211–33.

Osborne, Thomas (2004), 'On Mediators: Intellectuals and the Ideas Trae in the Knowledge Society', *Economy and Society*, 33 (4): 430–47.

Page, Will (n.d.), 'Rockonomics', *Amazon.co.uk*. Available online: http://www.amazon.co .uk/Rockonomics/lm/R3620VP4LF8L7L.

Parker, Martin 2013 'Beyond Justification: Dietrologic and the Sociology of Critique', in Paul Du Gay and Gareth Morgan (eds.), *New Spirits of Capitalism: Crises, Justifications and Dynamics*, 124–41, Oxford: Oxford University Press.

Parkinson, Tom, and Gareth Dylan Smith (2015), 'Towards an Epistemology of Authenticity in Higher Popular Music Education', *Action, Criticism and Theory for Music Education*, 14 (1): 93–127.

Parsons, Patrick R. (1988), 'The Changing Role of Women Executives in the Recording Industry', *Popular Music and Society*, 12 (4): 31–42.

Passman, Don (2014), *All You Need to Know About the Music Business*, 8th edn, New York: Simon & Schuster.

Peck, Jamie (2005), 'Struggling with the Creative Class', *International Journal of Urban and Regional Research*, 29 (4): 740–70.

Peterson, Richard A., and David G. Berger (1971), 'Entrepreneurship in Organizations: Evidence from the Popular Music Industry', *Administrative Science Quarterly*, 16 (1): 97–106.

Petersson Macintyre, Magdalena (2014), 'Commodifying Passion: The Fashion of Aesthetic Labour', *Journal of Cultural Economy*, 7 (1): 79–94.

Pini, Maria (2001), *Club Cultures and Female Subjectivity: The Move from Home to House*, London: Palgrave.

Plato (1998), *Phaedrus*, New York: Cornell University Press.

Pollock, Neil, and Robin Williams (2009), *Software and Organisations: The Biography of an Enterprise-wide System or How SAP Conquered the World*, London: Routledge.

Poundstone, William (2011), *Priceless: The Hidden Psychology of Value*, Oxford: Oneworld Publications.

Powers, Devon (2010), 'Strange Powers: The Branded Sensorium and the Intrigue of Musical Sound', in Melissa Aronczyk and Devon Powers (eds.), *Blowing Up the Brand: Critical Perspectives on Promotional Culture*, 285–306, New York: Peter Lang.

Powers, Devon (2012), 'Long-haired Freaky People Need to Apply: Rock Music, Cultural Intermediation, and the Rise of the "Company Freak"', *Journal of Consumer Culture*, 12 (1): 3–18.

Powers, Devon (2013), 'Now Hear This: The Promotion of Music', in Matt McAllister and Emily West (eds.), *The Routledge Companion to Advertising and Promotional Culture*, 313–25, New York: Routledge.

Pratt, A. C. (2004), 'Creative Clusters: Towards the Governance of the Creative Industries Production System?', *Media International Australia Incorporating Culture and Policy*, 112: 50–66.

Pratt, A. C. (2009), 'Urban Regeneration: From the Arts "Feel Good" Factor to the Cultural Economy: A Case Study of Hoxton, London', *Urban Studies*, 46 (5&6): 1041–61.

Pratt, A. C., and Toby Bennett (2022), *Everything You Always Wanted to Know About Data for the Cultural and Creative Sector Production System, But Were Afraid to Ask: Part 1 – Problems of Statistical Description*, CICERONE report D4.2, Amsterdam: University of Amsterdam.

Qiu, Jack Linchuan, Melissa Gregg, and Kate Crawford (2014), 'Circuits of Labour: A Labour Theory of the iPhone Era', *Triple C*, 12 (2): 564–81.

Ramsey, Phil, and Andrew White (2015), 'Art for Art's Sake? A Critique of the Instrumentalist Turn in the Teaching of Media and Communications in UK Universities', *International Journal of Cultural Policy*, 21 (1): 78–96.

Redhead, Steve (1997), *Subculture to Clubcultures: Introduction to a Popular Cultural Studies*, Oxford: Blackwell.

Redhead, Steve (2000), *Repetitive Beat Generation*, Edinburgh: Canongate.

Redmond, Steve (2017), 'What the Hell's Happening to Music's Trade Press? (and What Does it Mean for the Rest of Us?), *Music Business Worldwide*, 27 November. Available online: https://www.musicbusinessworldwide.com/hells-happening -musics-trade-press-mean-rest-us/ (accessed 9 February 2024).

Rekret, Paul (2019), 'Melodies Wander Around as Ghosts: On Playlist as Cultural Form', *Critical Quarterly*, 61: 2.

Reynolds, Simon (1998), *Energy Flash: A Journey Through Rave Music and Dance Culture*, London: Picador.

Reynolds, Simon (2011), *Retromania: Pop Culture's Addiction to its Own Past*, London: Faber.

Rifkin, Jeremy (2000), *The Age of Access: The New Culture of Hypercapitalism, Where all of Life is a Paid-for Experience*, New York: Penguin Putnam.

Robertson, Craig (2017), *The Filing Cabinet: A Vertical History of Information*, Minneapolis: University of Minnesota Press.

Rogers, Jim (2013), *The Death and Life of the Music Industry in the Digital Age*, London: Bloomsbury.

Roscoe, Philip (2015), *A Richer Life: How Economics Can Change the Way We Think and Feel*, London: Penguin.

Roscoe, Philip (2023), *How to Build a Stock Exchange*, Bristol: Bristol University Press.

Ross, Andrew (2003), *No Collar: The Humane Workplace and Its Hidden Costs*, New York: Basic Books.

Rossiter, Ned (2016), *Software, Infrastructure, Labor: A Media Theory of Logistical Nightmares*, London: Routledge.

Roy, Donald (1960), '"Banana Time": Job Satisfaction and Informal Interaction', *Human Organization*, 18 (4): 158–68.

Roy, Elodie A. (2015), *Media, Materiality and Memory: Grounding the Groove*, Aldershot: Ashgate.

Russell, Andrew, and Lee Vinsel (2016), 'Hail the Maintainers', *Aeon*, 7 April. Available online: https://aeon.co/essays/innovation-is-overvalued-maintenance-often-matters -more.

Sandoval, Marisol (2014), *From Corporate to Social Media: Critical Perspectives on Corporate Social Responsibility in Media and Communication Industries*, London: Routledge.

Sanneh, Kelefa (2004), 'The Rap Against Rockism', *The New York Times*, 31 October. Available online: https://www.nytimes.com/2004/10/31/arts/music/the-rap-against -rockism.html.

Sarpong, David, Shi Dong, and Gloria Appiah (2016), '"Vinyl Never Say Die": The Re-incarnation, Adoption and Diffusion of Retro-technologies', *Technological Forecasting & Social Change*, 103: 109–18.

Savage, Mike (2009), 'Against Epochalism: An Analysis of Conceptions of Change in British Sociology', *Cultural Sociology*, 3 (2): 217–38.

Sayer, Andrew (2000), 'Moral Economy and Political Economy', *Studies in Political Economy*, 61: 79–103.

Schmidt, Eric, and Jonathan Rosenberg (2014), *How Google Works*, London: John Murray.

Schwartz, Barry (2005), *The Paradox of Choice: Why More is Less*, New York: Harper.

Seaver, Nick (2022), *Computing Taste: Algorithms and the Makers of Music Recommendation*, Chicago: University of Chicago Press.

Sedgwick, Eve Kosofsky (1997), *Touching Feeling: Affect, Pedagogy, Performativity*, Durham: Duke University Press.

Sheridan, Jim (2003), *Pop Idol*, EDM1838A2, 5 November.

Siciliano, Michael (2016), 'Disappearing into the Object: Aesthetic Subjectivities and Organizational Control in Routine Cultural Work', *Organization Studies*, 37 (5): 687–708.

Siciliano, Michael (2021), *Creative Control: The Ambivalence of Work in the Culture Industries*, New York: Columbia University Press.

Sinnreich, Aram (2013), *The Piracy Crusade: How the Music Industry's War on Sharing Destroys Markets and Erodes Civil Liberties*, Amherst and Boston: University of Massachusetts Press.

Siskin, Clifford (2016), *System: The Shaping of Modern Knowledge*, Cambridge: MIT Press.

Skillset (2012), *CIC Skills Group: Report to Creative Industries Council*, London: Skillset.

Smirke, Richard (2011), 'BPI Establishes Fund for Indie Labels Impacted by London Riots', *Billboard*, 15 August. Available online: https://www.billboard.com/music /music-news/bpi-establishes-fund-for-indie-labels-impacted-by-london-riots -1175706/.

Smith, Chris (1998), *Creative Britain*, London: Faber.

Smith, Gareth Dylan, Zack Moir, Matt Brennan, Shara Rambarran, and Phil Kirkmann (2017), *The Routledge Research Companion to Popular Music Education*, London: Routledge.

Smith Maguire, Jennifer, and Julian Matthews (2012), 'Are We All Cultural Intermediaries Now? An Introduction to Cultural Intermediaries in Context', *European Journal of Cultural Studies*, 15: 5.

Southall, Brian (2009), *The Rise and Fall of EMI Records*, London: Omnibus.

Spitz, David, and Starling D. Hunter (2005), 'Contested Codes: The Social Construction of Napster', *The Information Society*, 21 (3): 169–80.

Stahl, Matt (2013), *Unfree Masters: Popular Music and the Politics of Work*, Durham: Duke University Press.

Stahl, Matt, and Leslie M. Meier (2012), 'The Firm Foundation of Organizational Flexibility: The 360 Contract in the Digitalizing Music Industry', *Canadian Journal of Communication*, 37: 441–58.

Sterne, Jonathan (2012), *MP3: The Meaning of a Format*, Durham and London: Duke University Press.

Sterne, Jonathan (2014), 'There Is No Music Industry', *Media Industries*, 1 (1): 50–5.

Steward, Sue, and Sheryl Garrett (1984), *Signed, Sealed & Delivered: True Life Stories of Women in Pop*, London: Pluto Press.

Stiegler, Bernard (2014), *The Lost Spirit of Capitalism: Disbelief and Discredit, 3* (Cambridge: Polity)

Stratton, Jon (1982), 'Reconciling Contradiction: The Role of the Artist and Repertoire Person in the British Music Industry', *Popular Music & Society*, 8 (2): 90–100.

Straw, Will (2011), 'Music and Material Culture', in Martin Clayton, Trevor Herbert, and Richard Middleton (eds.), *The Cultural Study of Music*, 227–36, New York: Routledge.

Street, John, Dave Laing, and Simone Schroff (2018), 'Regulating for Creativity and Cultural Diversity: The Case of Collective Management Organisations and the Music Industry', *International Journal of Cultural Policy*, 24 (3): 368–86.

Sum, Ngai-Ling, and Bob Jessop (2013), *Towards a Cultural Political Economy: Putting Culture in its Place in Political Economy*, Cheltenham: Edward Elgar Press.

Sun, Hyojung (2019), *Digital Revolution Tamed: The Case of the Recording Industry*, London: Palgrave.

Tagg, Philip (1999), 'Understanding Musical Time-sense'. Available online: https://www.tagg.org/articles/timesens.html.

Tarnoff, Ben, and Moira Weigel (2020), *Voices from the Valley: Tech Workers Talk About What They Do—And How They Do It*, New York: FSG Originals x Logic.

Taylor, Mark, and Dave O'Brien (2017), '"Culture Is a Meritocracy": Why Creative Workers' Attitudes May Reinforce Social Inequality', *Sociological Research Online*, 22 (4): 27–47.

Taylor, Timothy D. (2012), *The Sounds of Capitalism*, Chicago: University of Chicago Press.

Taylor, Timothy D. (2013), 'Music in the New Capitalism', in Angharad N. Valdivia and Vicki Mayer (eds.), *The International Encyclopedia of Media Studies, Vol.II: Media Production*, 151–70, London: Blackwell.

Tholen, Gerbrand (2017), *Graduate Work: Skills, Credentials, Careers, and Labour Markets*, Oxford: Oxford University Press.

Thompson, Paul, and Mats Alvesson (2004), 'Bureaucracy at Work: Misunderstandings and Mixed Blessings', in Paul du Gay (ed.), *The Values of Bureaucracy*, 89–114, Oxford: Oxford University Press.

Thornton, Sarah (1995), *Club Cultures: Music, Media and Subcultural Capital*, Cambridge: Blackwell.

Thornton, Sarah (2003), 'An Academic Alice in Adland: Ethnography and the Commercial World', *Critical Quarterly*, 41 (1): 58–68.

Thornton, Patricia H., William Ocasio, and Michael Lounsbury (2015), 'The Institutional Logics Perspective', in Robert A. Scott and Marlis C. Buchmann (eds.), *Emerging Trends in the Social and Behavioural Sciences: An Interdisciplinary, Searchable and Linkable Resource*. https://doi.org/10.1002/9781118900772.etrds0187.

Thrift, Nigel (2005), *Knowing Capitalism*, London: Sage.

Thrift, Nigel (2008), 'The Material Practices of Glamour', *Journal of Cultural Economy*, 1: 1.

Titlow, John Paul (2014), 'How Spotify's Music – Obsessed Culture Keeps Employees Hooked', *Fast Company*. Available online: http://www.fastcompany.com/3034617/how-spotifys-music-obsessed-culture-makes-the-company-rock.

Tooze, Adam (2019), *Crashed: How a Decade of Financial Crises Changed the World*, London: Penguin.

Towse, Ruth (2006), 'Human Capital and Artists' Labour Markets', in Victor Ginsburgh and David Throsby (eds.), *Handbook of the Economics of the Arts and Culture*, 865–94, Amsterdam : North Holland Elsevier.

Toynbee, Jason (2000), *Making Popular Music*, London: Arnold.

Tschmuck, Peter (2009), 'Copyright, Contracts and Music Production', *Information, Communication & Society*, 12 (2): 251–66.

Tunstall, Jeremy (2000), *Media Occupations and Professions: A Reader*, Oxford: Oxford University Press.

Turner, Fred (2006), *From Counterculture to Cyberculture: Stewart Brand, the Whole Earth Network, and the Rise of Digital Utopianism*, Chicago: University of Chicago Press.

Turner, Fred (2008), 'Millenarian Tinkering: The Puritan Roots of The Maker Ethic', *Technology and Culture*, 59 (4): S160–82.

UK Music (n.d.), 'Equality & Diversity', UK Music. Available online: https://www.ukmusic.org/equality-diversity/ (accessed 12 January 2018).

UK Music (2013), *The Economic Contribution of the Core UK Music Industry*, London: UK Music. Available online: http://www.ukmusic.org/assets/general/Summary_Document_-_The_Economic_Contribution_of_the_Core_UK_Music_Industry.pdf.

UK Music (2014), *Internship Code of Practice*, London: UK Music.

UK Music (2017), *Measuring Music: 2017 Report*, London: UK Music.

UMG and Music Week (n.d.), *Everything You Need to Know, to Get A Job In Music*, London: Music Week.

Vats, Anjali (2019), 'Prince of Intellectual Property: On Creatorship, Ownership, and Black Capitalism in Purple Afterworlds (Prince in/as Blackness)', *Howard Journal of Communications*, 30 (2): 114–28.

Vinnicombe, Chris (2011), 'UK Indie Labels Hit by Sony Warehouse Fire', *MusicRadar*, 9 August. Available online: https://www.musicradar.com/news/guitars/uk-indie -labels-hit-by-sony-warehouse-fire-487185.

Vonderau, Patrick (2019), 'The Spotify Effect: Digital Distribution and Financial Growth', *Television & New Media*, 20 (1): 3–19.

Wald, Melissa (2011), 'Music Industry Administration in the Digital Age – A Brief Description of the Evolution of Current Industry Practices and Some of the Challenges to Come: Will Our College Graduates Possess the Necessary Skills to Enter this Marketplace?', *Journal of the Music & Entertainment Industry Educators Association*, 11 (1): 223–38.

Wall, Tim (2013), 'The X Factor', in Peter Bennett and Julian McDougall (eds.), *Mythologies Today: Barthes Reimagined*, 19–23, London: Routledge.

Warde, Allan, David Wright, and Modesto Gayo-Cal (2007), 'Understanding Cultural Omnivorousness: Or, the Myth of the Cultural Omnivore', *Cultural Sociology*, 1 (2): 143–64.

Wardle, Ben (2008a), 'Inside the Record Company Cauldron…', *A&Rmchair*, 14 June. Available online: http://benwardle.blogspot.com/2008/06/inside-record-company -cauldron.html.

Wardle, Ben (2008b) 'A&Rs Are the Unsung Heroes of the Music Industry', *The Guardian*, 3 March. Available online: https://www.theguardian.com/music/ musicblog/2008/mar/03/arsaretheunsungheroesoft.

Warner, Simon (2017), 'Where to Now? The Current and Future Trajectory of Popular Music Studies in British Universities', in Gareth Dylan Smith, Zack Moir, Matt Brennan, Shara Rambarran, and Phil Kirkmann (eds.), *The Routledge Research Companion to Popular Music Education*, 127–38, London: Routledge.

Watson, Allan (2008), 'Global Music City: Knowledge and Geographical Proximity in London's Recorded Music Industry', *Area*, 40 (1): 12–23.

Watson, Allan (2012), *Cultural Production in and Beyond the Recording Studio*, London: Routledge.

Watson, Allan, and Andrew Leyshon (2022), 'Negotiating Platformisation: MusicTech, Intellectual Property Rights and Third Wave Platform Remediation in the Music Industry', *Journal of Cultural Economy*, 15 (3): 326–43.

Weber, Max (2001), *The Protestant Ethic and the Spirit of Capitalism*, London: Routledge.

Weston, Kris (2014), 'Orb Rant', *KrisWeston.com*. Archived by Wayback Machine, captured on 16 June, 2015. Available online: https://web.archive.org/web /20150616151158/https://krisweston.com/articles/orbrant/.

Wheeldon, Jonathan (2009), *Struggling to Define Value: A Critical and Discourse-Based Study of Strategic Sensemaking in the Recorded Music Industry*, Unpublished DBA thesis, Reading: Henley Business School.

Wheeldon, Jonathan (2014), *Patrons, Curators, Inventors and Thieves: The Storytelling Contest of the Cultural Industries in the Digital Age*, London: Palgrave Macmillan.

Whelan, Andrew (2014), 'The Morality of the Social in Critical Accounts of Popular Music', *Sociological Research Online*, 19 (2): 1–11.

Wikström, Patrik (2012), 'A Typology of Music Distribution Models', *International Journal of Music Business Research*, 1 (1): 7–20.

Wikström, Patrik (2013), *The Music Industry: Music in the Cloud*, 2nd edn, Cambridge: Polity Press.

Wikström, Patrik, and Robert DeFilippi (2016), *Business Innovation and Disruption in the Music Industry*, Cheltenham: Edward Elgar.

Wilkinson, Kenton, and Patrick Merle (2013), 'The Merits and Challenges of Using Business Press and Trade Journal Reports in Academic Research on Media Industries', *Communication, Culture & Critique*, 6: 415–31.

Williamson, John, and Martin Cloonan (2007), 'Rethinking the Music Industry', *Popular Music*, 26 (2): 305–22.

Williamson, John, and Martin Cloonan (2013), 'Contextualising the Contemporary Recording Industry', in Lee Marshall (ed.), *The International Recording Industries*, 11–30, London: Routledge.

Williamson, John, and Martin Cloonan (2016), *Players' Work Time: A History of the British Musicians' Union, 1893–2013*, Manchester: Manchester University Press.

Williamson, John, Martin Cloonan, and Simon Frith (2011), 'Having an Impact? Academics, the Music Industries and the Problem of Knowledge', *International Journal of Cultural Policy*, 17 (5): 459–74.

Winter, Carsten (2012), 'How Media Prosumers Contribute to Social Innovation in Today's New Networked Music Culture and Economy', *International Journal of Music Business Research*, 1 (2): 46–73.

Witt, Stephen (2015), *How Music Got Free: The Inventor, the Mogul and the Thief*, London: Vintage.

Work Spy (2014a), 'Work Spy', *Elle*, June 2014, 89–90.

Work Spy (2014b), 'Work Spy', *Elle*, August 2014, 110–11.

Yetnikoff, Walter (2004), *Howling at the Moon*, New York: Broadway Books.

Young, Liam Cole (2017), *List Cultures: Knowledge and Poetics from Mesopotamia to Buzzfeed*, Amsterdam: Amsterdam University Press.

Index

1960s, the
  and countercultural movements   10,
    13, 20, 192
  as evocative of style and
    genre   101, 117
  house hippies and company
    freaks   20, 192–3, 214
  recruitment of senior executives
    in   42, 67
1990s, the
  Britpop in   10, 69
  changes in knowledge production
    in   35, 48, 158–9
  corporate culture and practice
    in   65, 91
  dance music subcultures in   69–70,
    99, 130
  emergence of creative economy policy
    in   21, 40, 69
  as formative period for contemporary
    executives   67–70
  industrial change in   21, 57–8, 120,
    122, 125, 127, 166,   191–2
  in *Kill Your Friends*   79–80, 84–6, 93
    (*see also Kill Your Friends*)
  as period of music industry boom and
    excess   2, 79–80
  Prince's contractual dispute in   37
1999
  as origin of the *Idol* franchise   212
  partying like it's   34, 36–7
  representations of in popular
    music   34
  as symbolic starting point of the
    study   1–2, 4, 33–5, 50–1,
    57–8, 60
  and the Y2K millennium bug   34
2000s, the
  audience measurement in   109
  disavowal of file-sharing   107
  fluidity metaphors in   31

industry change in   1–2, 13, 57–8,
    119, 127, 131, 147–8, 171
  organisational restructuring in   60,
    89, 139
  poor industry reputation in   80
  studies of cultural work in   22

artists and repertoire (A&R)
  conflicts with   92, 146, 156–7
  the culture of   79–80, 102, 190, 198–9
  as object of study   14, 179
  as organisational division   63, 64, 66,
    123, 144, 161
  representations of   82, 84–5, 93
Abbey Road studios   65
access
  as 'epochal' keyword   4, 30–3, 35,
    50–2, 119
  to catalogues of intellectual property
    rights   29, 202
  to knowledge and information   44–
    50, 90, 132, 134, 136–7, 139–40,
    155–6, 175, 188
  to music and media content   2–3, 14,
    36–40, 105
  organisational systems as means of
    controlling   129, 134, 145
  to work and employment   40–4,
    69–70, 152
Access Industries   58
accountancy
  as a profession   128, 147, 150, 155,
    157, 162, 187
  systems and procedures for
    managing   25, 120–3, 127, 129
active audiences   47, 66, 83, 92, 94,
    107–11, 115
administration
  discourses of   134–5
  growth in   24, 128, 150, 155
  identity of   74, 143–4

of legal and financial duties   29, 61,
        140, 183
organisational function and logic
        63–4, 75, 127, 132
        (*see also* enterprise)
public   55
workers   3, 24–6, 97, 101, 131–3,
        144–9, 201–3
Adorno, Theodor   17–19, 47, 103. *See
        also* Critical Theory (Frankfurt
        School)
aesthetics
    innovation and change in   22, 28–9,
        48, 102
    and the judgment of value   13, 20, 26,
        66, 68, 110, 112, 115, 187, 201
        (*see also* valuation)
    of office life   96–101, 168
    of popular music   23, 34, 38, 69, 201
affect   38, 49–50, 88–9, 106–8, 113–16,
        194. *See also* passion
affective economics   83, 108
age. *See also* youthfulness
    of books and their subject
        matter   179–80
    in education, perceptions of   161–2
    of executives   47
    of institutions   161, 183
    of musicians   103
    of organisational functions   91
Ahmed, Sara   93–4, 113
algorithmic recommendation   49–50,
        103, 110, 115, 195
amateurism   9–10, 13, 51, 60, 112, 113
Amazon   162, 175–6
Anderson, Chris   10, 49, 108
anthropology
    as academic discipline   65, 135
    and ethnographic method   4, 46–7,
        110, 176, 191–2
    as lay knowledge   32–3, 110, 194, 198
Apple   10, 11, 119, 126, 160, 173
apprenticeship   41, 44, 154, 187
archiving   63, 101, 131, 133, 135–42,
        145–6, 149, 193. *See also*
        database; warehousing
art-commerce tensions   15, 25, 63–4,
        115, 132

Artificial Intelligence (AI)   202, 213
artistic critique. *See under* critique
art worlds (Howard Becker)   25, 27,
        47, 65
assets
    digital   134, 145 (*see also* database)
    financial   28–9, 61, 179 202 (*see also*
        capitalism, finance)
Atlantic Records   42, 111
audiences
    active   47, 83, 92, 109
    mass   82–3, 88, 92, 109, 193
autobiography. *See* biography
automation   36, 118, 127, 129, 131, 137,
        149, 213 n.1
avidity   110–13, 115. *See also* passion

back office   63, 97–9, 119–20, 123–5,
        127–9, 132–3, 139–48, 197–203
backstage   3, 30, 33, 88
Bangs, Lester   79, 179
Barfe, Louis   2, 30, 42, 79, 212 n.2
BBC, The   48, 54, 109, 117–18, 173, 199,
        214 n.2
Beatles, The   9–12, 20, 53, 69, 177
Becker, Howard   25, 27, 47, 65
Beese, Darcus   42, 55
Bertelsmann Music Group (BMG)   54,
        57–8, 65, 171
Big Music   53–77, 192, 194
Billboard (trade journal). *See under* trade
        journals
biography
    executive   42, 47, 95, 177, 179, 182
    interviewees' appeal to their own   70–
        4, 106–7, 111–12
    organisational   32, 124
    recommendations of   177–8, 180, 184
    as source of legitimacy   87, 169–70
    this author's   51
Black music, historic importance to
        industry   11, 54–5, 64, 200, 202
Blackwell, Chris   42–3, 64
blockchain   129, 202, 213 n.1
Boltanski and Chiapello (*The New Spirit of
        Capitalism*)   4, 12, 16–21, 50,
        176, 194, 196–7
boredom   19, 111, 117, 143–4, 148

Born, Georgina   27, 48, 50, 59, 66, 102
Bourdieu, Pierre   20, 26, 65, 67, 185, 187, 192
Bowie, David   31
Britishness   1, 11, 40–1, 53–5, 64–5, 84, 86
British Phonographic Industry (BPI)   9, 29–31, 37, 44–5, 59, 107, 153, 155, 159, 192, 211 n.1
Britpop   10, 68, 69, 200
bureaucracy
  identification with   42, 72, 149, 203
  as mode of organisation   63, 65, 120, 124, 194
  opposition to   26, 35, 97, 131–2, 151, 163
  technologies of   137–41
business affairs   24, 26, 71, 140, 144–5
business models   4, 10, 22, 31–2, 51, 80, 119, 126, 155, 183
Business School   75, 157, 160, 162, 184, 186, 188

Caldwell, John   47–8, 65–6, 174–5, 177, 193
California   10, 21, 35, 65, 154, 192
"Californian Ideology" (Barbrook and Cameron)   10
call centres   131, 148
capitalism
  consumer   17–19, 21
  corporate   56–8, 60–2, 131
  financial   16, 28–9, 183, 211–12 n.1
  gentlemanly   65, 67, 75
  knowing (Nigel Thrift)   75, 153, 170, 182–3
  musical (Georgina Born)   13, 27, 59, 67, 75, 93, 152, 183–4
  neoliberal   16, 22
  post-industrial   21, 55, 131
Capitol Records   64
careers advice   2–3, 5, 41, 44, 154, 163–4
cassette tapes. *See under* formats
catalogue
  management of   129, 141, 145
  number   121, 133
  as organisational division   4, 63–4, 103–4, 117

of rights and repertoire   36, 122, 124, 136, 202
strategic importance of   4, 29, 40, 116–17
users' attempts to   36–7, 138
charts   104, 122, 171, 173, 212 n.1
Chess Records   11
Chiapello, Ève   179. *See also* Boltanski and Chiapello
club
  cultures   69, 71, 86, 99, 156, 160, 192
    (*see also* subculture)
  goods   39
  old boys'   53, 57, 67
  private members'   70, 75, 191
cluster, industry   11, 53–5, 74–5
collective management organisations   14, 58, 61, 119, 153, 213 n.1.
  See also PRS
commodity, music as a   36, 46, 82, 103, 112–13
communications, corporate   45, 59, 61, 81, 87, 89, 91–2, 94, 100, 105, 195, 198
communicative action   3, 91, 116, 133
community, of music professionals   56, 59–66, 96, 101, 113, 116–17, 171–5, 177, 190, 195, 203
consumer electronics   34, 36, 58, 61, 119, 122
consumers (of industry
    commentary)   172–8, 181–2, 199
consumers (of music)
  attempts to think like   32, 73–4, 90–1, 94, 110–11, 194–5
  employees as   66–9, 104–5, 111–18
  opinions of music industry from   1–2, 14, 81–8, 91–4, 104
  tastes, preferences, behaviours and habits of   31–2, 63, 83, 105, 107–8, 120, 181, 183, 197
consumption
  conspicuous   88, 108
  ethics of   2, 10, 15–16, 56–7, 106–8, 195–6
  industrial management of   18, 21, 35–40, 87, 119–23

as object of scholarly analysis   46–7, 51–2
contracts
    employment   21, 32, 46, 112, 124–5, 128–9, 132
    as organisational device   62–4, 122–3, 129, 133–7, 140–4, 148, 186
    recording   23, 39, 47, 82, 125–6, 130–2
    smart   213 n.1
    supplier   23, 62, 125–6, 128
conversion   96, 105–11. *See also* religious faith
cool
    Britannia   69
    employees   72, 102, 117 (*see also* geeks)
copyright
    as basis for technical framework   37–9, 61, 120–1, 131–5, 142
    control and ownership of
    creation of   21, 40
    economics of   37–8
    exploitation of   37, 201–2
    legal frameworks of   10, 136
    moral economy of   2–3, 14, 24–6, 202
    as object of public policy and lobbying   3, 13–14, 33, 41, 45, 59, 62, 156, 173, 195
    opposition to   10, 107
    protection and enforcement   16, 37–8, 122
Corporate Social Responsibility (CSR)   15, 89–90
countercultures   18–21, 42, 67, 108
Covid-19 pandemic   20, 74, 200
Cowell, Simon   42, 81–4, 87, 212 n.3
craft skills, appeals to   18, 25, 92, 115, 144, 158, 196
Creation Records   138, 180
creative class   21, 29, 55, 67, 69–70, 154
creative industries (policy)   21–2, 33, 40–1, 44, 69, 153
crisis
    capitalist   10, 19, 22–3, 56
    legitimation (Jürgen Habermas)   3
    makes the hidden rules of the system visible   81

management   187
mental health   29
music industry in   3, 29, 30, 60, 80
critical theory (Frankfurt School)   3, 17–19, 47, 103
criticism
    academic   24, 26, 46, 70, 190, 193, 203
    among audiences   81–4
    of digital platforms   9–10
    of industry from the "inside"   86–7, 103, 197, 199–201
    of industry from the "outside"   2, 198
    music and cultural   14, 28–9, 68, 82–3, 87, 138, 159
    popular and everyday   67, 115, 161, 175–6, 192 (*see also* evaluative judgment)
    in popular music studies   12–15, 158–60
    unwanted   26
critique. *See also* Boltanski and Chiapello
    artistic   19–20, 198, 200
    social   12, 17–19, 24, 48
cultural economy perspective   48–9, 66
cultural industries   21, 24–5, 29, 32, 47, 69, 166
cultural intermediaries   14, 20, 23, 26, 29, 68
cultural policy   40–1
cultures
    corporate and organisational   48, 64–5, 80, 95–105, 111, 116–18, 123, 129, 192
    industry production (Caldwell)   5, 48, 57–8, 65–70, 75, 80–1, 103, 133, 174, 199, 201
    music   33, 42, 75, 99–102, 118, 195
Curran, James   35
custodianship   12, 165, 187

data. *See also* metadata
    'big'   49, 186
    statistical   44–5, 66, 91, 110, 169–70, 177, 180–1, 194, 197, 199–200
database   27, 28, 90, 122–3, 128, 131, 134–7, 140–2, 145, 158
Def Jam   42
deindustrialisation   19–21, 55, 131

Demon Music Group   214 n.2
deskilling   144, 213 n.1
Digital Economy Act (2010)   13
Digital Rights Management   38–9
disciplines   91, 146, 157–60, 170, 182–8
disruption   2, 27, 30–5, 50, 92, 119–20,
      147–8, 172, 180, 184–5, 202
distribution. *See* infrastructure
diversity and inequity   22, 44, 69, 74–5,
      90, 154, 198–200
division of labour   18, 24–5, 96, 134
downloads   2, 31, 36, 38, 134, 141, 155
Duggan, Mark   56

economics
   affective (Henry Jenkins)   83
   cultural   154
   influence of   16, 169, 181
   of information   39, 183
   profession   175–6, 182–3, 185
   textbooks   177–8
   theory   16, 162, 181
education
   school   65, 67, 71, 90, 154–5
   system   41, 152, 158
   university   41, 67–8, 150–67,
      182, 192
elites   40, 65–7, 70, 75, 86, 90, 156–7,
      169, 171, 176
EMI   33, 42, 53, 57–8, 64–5, 133, 156–7,
      173, 180
engineering
   profession   18, 186
   software   110, 115, 155
   sound   14–15, 61, 62, 65
enterprise
   as organisational logic   32, 62–3, 124,
      132, 170
   resource planning (ERP)   120,
      127–9, 137
epochalism   4, 35, 51, 91, 103
Epstein, Brian   12
Ertegun, Ahmet   42, 84, 111
eschatology   35. *See also* religious faith
ethics   10, 45, 111, 203
ethnography. *See under* anthropology
European Union (EU)   10, 64, 124,
      128, 160

evaluative judgment   66, 79–94, 103, 115,
      123, 161, 170
excess, performances of   19, 79–81, 99,
      178–9, 196
extraction
   industrial   13, 59
   of information   48–9, 138
   of value   13, 26, 31, 46, 200, 202

Factory Records   121
factory work   13, 143
fandom   102, 105–6, 111, 115
fashion   40, 48, 54, 61, 70, 74, 97–8,
      100–2, 109
feminism   20, 70, 75, 198–9, 203
filesharing   10, 33, 156
filing   133, 137–41, 144, 146, 149, 195
financialisation. *See under* capitalism
Fiske, John   47, 83
Florida, Richard   21, 55, 67, 154–5
Forde, Eamonn   2, 14, 86, 157, 172,
      175–82
Fordism   17–20, 35
format
   cassette tape   31, 36–9
   MP3   33, 36, 38–9, 120–1, 135
   streaming media   2, 9–11, 29, 30, 36,
      39, 50, 103–4, 107, 109–11, 117,
      119, 125–6, 169, 200
   vinyl   30, 36, 38–9, 138
Fraser, Nancy   20, 201
Frith, Simon   14, 23, 33, 38, 48, 67–8,
      159, 192
frontline (organisational division)   63–4,
      104, 117, 144–6, 202
Fuller, Simon   82, 84

geeks   102, 110, 117. *See also* cool
Geffen, David   42–3, 164
gender   13, 18–20, 69–70, 73–4, 108,
      163–4, 198–9
Generation X   66–9, 99. *See also*
      intergenerational conflict
genre
   discourse   116
   literary   86–7, 176–8, 181–2, 184
   music   28, 48, 84, 101–2, 114–15, 129,
      135, 168, 176, 199

gentrification 55
geography 53–7, 62, 74–5, 127–8
Giddens, Anthony 49, 170
Gill, Rosalind 22, 69–71, 198, 199
glamour, misconceptions of 3, 11, 27,
    42–3, 51, 87–8, 123, 196
globalisation 21, 48, 54, 65, 120, 127
gonzo 79–80, 85–6, 179, 211 n.1
Goodman, Fred 2, 30
Google 74, 126, 160, 162, 183
gossip 3, 47, 58, 66, 86, 100, 104, 116,
    173–5, 195
graduate training schemes 43–4, 150–2
Grainge, Lucian 9, 11, 13, 29, 125
Grateful Dead, The 10, 21
Guardian, The 9–11, 156

Habermas, Jürgen 3. *See also* critical
    theory (Frankfurt School)
Hammersmith (West London) 53–4
Hands, Guy 156
Harris, John 9–14, 28, 42
haunting 28–9, 34, 79
Hayes (West London) 20, 54
headphones 38, 112, 117, 142
Hebdige, Dick 47, 54–5, 71, 192
Hennion, Antoine 14, 38, 113
heritage
    acts 103, 172–3, 200
    awareness of 102, 115, 181
    displays of 97, 100, 103, 168
    economy 40
    as source of identity 54, 64, 74
Hesmondhalgh, David 15, 21, 24–5, 36,
    38, 47–8, 125, 138–9, 200
hidden abode of production 46, 96. *See
    also* Marx, Karl
hierarchy
    of needs in Maslow 110
    organisational 63–4, 75, 97–8, 117,
        123–4, 129, 132–4, 144–9
    social, critique of 18, 48, 98, 139
hipsters 28, 71, 98, 99. *See also* creative
    class; scenesters
Hirschmann, Albert 17, 93, 197
Hoxton (East London) 70
human resources (HR) 24, 43, 45, 63–4,
    66, 97, 127, 188, 197

Idol Shows 81–5, 87–8, 93, 173, 212 n.3
inclusion. *See* diversity and inequity
independent
    ethics 12–13, 22
    labels 21, 42, 56, 130, 135, 139, 201
    sector, representation of 59–
        60, 153–4
information retrieval 131, 134–7
infrastructure
    aesthetic experience of 112
    as analytic perspective 27, 49–50, 59
    digital 5, 36, 60–1, 120, 134–5
    maintenance of 149, 151
    organisational 103, 127, 131,
        138, 148–9
    physical 56, 58, 121, 142, 144 (*see
        also* warehousing)
innovation
    discourses of 35, 41, 63, 102–3,
        154, 169
    disruptive (*see* disruption)
    and infrastructure 36, 119, 195
    organisational 64, 138, 185
    processes of 26, 124, 153, 170
insight. *See* market research
institutional logics 62–4
intellectual property (IP). *See also*
        copyright
    economics of 37–8
    and knowledge 175
    and physical property 54
intellectuals 67, 70, 165, 170, 176, 181,
        185, 192–3, 196, 203. See also
        thought leaders
intergenerational conflict. *See also* age;
        Generation X; millennials;
        youthfulness
    and social change 18
    in the workforce 5, 91–3, 99, 102,
        117, 184–5
intermediation
    business 2, 26, 40, 61, 202
    cultural 14, 20, 23, 26, 29, 68, 185
    and disintermediation 2, 119, 129,
        213 n.1
    of knowledge and information 170–
        2, 176
    and mediation 170

International Federation of Phonographic Industries (IFPI)   2, 31, 44–5, 122, 134, 169
International Standard Recording Code (ISRC)   61, 121–2, 137
internships   24, 43–4, 68, 100, 140, 163, 198
Island Records   42, 54, 55, 64, 130
iTunes   11, 119

Jenkins, Henry   47, 83
Jewish music cultures, historic importance to industry   11–12
journalism
    investigative   46, 178–9
    music (*see under* criticism, music and cultural)
    trade   58, 112, 170–4, 175, 182, 192 (*see also* trade journals)

Keen, Andrew   10
Kelly, Kevin   10, 108
Kensington (West London)   53–5, 64, 66, 74–5
Kentish, Joe   55
Kill Your Friends (John Niven)   79–81, 84–7, 93, 179
Korczynski, Marek   103, 143

label copy   145. *See also* liner notes
labour process (theory)   46, 138, 140
labour union. *See* trade union
Latour, Bruno   49, 112–13, 149, 193
leadership   3, 44, 48, 55, 155, 196, 200, 202
legitimacy   9, 14, 16, 28, 33, 60, 72–3, 81, 87, 89, 105, 107, 110–11, 119, 122, 130, 157, 159, 162, 193–5, 202
    cultural   9, 72–3, 81, 202
    of digital music services   33, 105–7, 119, 122
    disciplinary   110, 159
    economic   16, 28
    organisational   3–4, 14, 60, 89, 111, 194
    professional   87, 157, 162, 193
Lepore, Jill   35, 180

Leyshon, Andrew   2–3, 14, 29–31, 37, 40, 56, 62, 108, 157
libertarianism   10
licensing
    importance of   4, 10, 14, 61, 62, 103, 119, 125–6, 141, 157, 169, 186, 202
    synch   29, 33, 40, 104, 129, 135
lifestyles   32, 70, 74, 99, 108–9, 130, 168
liner notes   27, 38, 111–13
listening   29, 36, 39, 83, 92–4, 103, 110, 112, 113, 119, 186, 189, 214 n.5
live music   45, 54, 58–62, 74, 199
logistics. *See* infrastructure
London Records   79
Long Tail, The (Chris Anderson)   108
Loog Oldham, Andrew   12, 164, 180, 182

McGee, Alan   138
McRobbie, Angela   22, 48, 69, 71
maintenance
    meaningful   5, 129, 147–9, 194–6, 198, 202–3
    work   26–7, 59, 120, 122–3, 145–7, 213 n.1
The Man   13, 18–19, 37
Marcuse, Herbert   19. *See also* Critical Theory (Frankfurt School)
marketing
    campaigns   83, 104–5, 115–16, 134, 141 (*see also* release)
    as the construction of markets   84, 169
    as organisational division   63, 66, 68, 102, 117, 128
market research
    and consumer insight   22–3, 89–93, 105, 108–10, 169–70
    and consumer segmentation   70, 108–10
    and cultural intermediaries   20, 23, 192
    as organisational division   89–91, 184–6
Marxist perspectives   17–18, 46–7, 49, 96, 112, 192

Master of Business Administration
(MBA). *See* Business school
masters, owning your 38, 202
material cultures 38, 97, 138
media studies 22, 150, 180
mediators. *See under* intermediaries
Meindl, Hugh 42–3
Mercury Records 64
mergers and acquisitions 33, 42, 54,
57–8, 97, 124–5, 127, 139, 173
meritocracy 21, 41, 43, 67, 69, 198–9.
See also diversity and inequity
millenarianism 33, 35. *See also*
religious faith
millennials 75, 81. *See also*
intergenerational conflict
misogyny 85, 131, 198. See also gender
Moby 33
moral economy 12, 15–16, 56–7, 108, 202
Morris, Meaghan 47
MP3. *See under* formats
MTV 97, 180
Muniesa, Fabian 83, 188, 193
Music Ally. See under trade press
Music Business Worldwide. See under
trade press
Musicians' Union (MU) 23, 197, 200.
See also trade unions
musicology (academic discipline) 48, 159
Music Week. See under trade press

Napier-Bell, Simon 2, 42, 179
Napster 33, 180, 182
negotiation 23, 45, 47, 61, 63, 101,
115, 125–6, 147, 151, 188–90,
200, 202
Negroponte, Nicholas 10
Negus, Keith 24, 26, 33, 48–9, 64–5,
67, 80–1, 119, 123, 127, 133,
179–82, 191
neoliberalism. *See under* capitalism
New Labour (government) 21, 33, 40–1
new materialism 49
New Spirit of Capitalism, The. See
Boltanski and Chiapello
Niven, John. *See Kill Your Friends*
Nixon, Sean 48, 70, 86, 97
Nostalgia 28, 34, 65, 69, 106

Notting Hill (West London) 54–5

Oakley, Kate 21, 57, 71
old boys' club 53, 57, 67
Orb, The 130
Orbison, Roy 13, 18–19

Page, Will 175–6, 179, 182–3
Pandora 126
Parlophone Records 54
passion
as emotional attachment 12, 15, 50,
68, 85, 88–9, 92, 100, 111–13
and interests 17, 63, 93–4, 197
for music 14, 51, 53, 66, 69, 70, 96,
99, 110, 112–14, 129, 133, 137,
149, 202
for music industry 73–4, 106–7,
138, 165
and paranoia (Deleuze and
Guattari) 46, 49, 189
zones 110
passionalization 5, 194–7, 202
Performing Right Society (PRS) 120,
153, 160, 182–3, 212–13 n.1
perks of the job, diminishing 56, 81,
88, 132
Peterson, Richard 14, 47
philosophy
continental 32, 49–50
explanations developed by 49
getting into chill mode 50
platonic 85, 93
political 17, 189
phones
hacking scandal 45
iPhone 10
mobile/smart 36, 141, 168, 211 n.1
office 65, 101, 131, 142, 146, 149
piracy 10, 37–9, 51, 92, 96, 106–8, 122,
169, 173
platforms, digital
marketing 83
organisational 127, 129
streaming 10–11, 29, 31, 39–40, 58,
61, 107, 119, 123, 125, 182–3, 200
playlists 28, 100, 103–5, 117–18, 138,
143, 193

policy, public   15, 21–2, 24, 33, 35, 40–1,
         50, 59, 61–2, 151, 153–4, 195,
         200, 202
Polydor Records   54, 64
polygram   33, 58, 130
popular music studies (PMS)   12, 14–15,
         23, 33, 47–8, 158–9, 166, 182,
         192, 213 n.1
Post-Fordism   22, 35, 48–9
postmodernism   35, 191
post room   42–3, 193
Pratt, Andy   22, 38, 44, 55, 70
precarity
    organisational   56
    in the workforce   22, 43, 46, 133, 148
Prince2   187, 214 n.4
Prince (Rogers Nelson), the artist formerly
         known as   34, 36, 37, 202
product codes   120–3, 129, 135, 193
production cultures. *See under* cultures
project management   127–8, 141, 187,
         199. *See also* Prince2
promotion. *See* marketing
property
    humans as (*see* slavery, resonances of
         in music industry)
    intellectual (*see* intellectual property)
    physical (*see* real estate)
psy-disciplines   19, 93, 97, 110, 143, 181,
         188–90
publishers
    book   121, 177, 179
    music   11, 41, 123

racism and racialization   11, 18, 37,
         55, 69, 86, 193, 199, 212 n.4,
         214 n.1. *See also* Black music,
         historic importance to industry
real estate   28, 53–6, 58, 61, 75, 139
realism
    and passion in work   114–16
    speculative   49
'Real world', The   115, 152, 161–6, 185–6
record men   9–13, 27, 29, 41–3, 47, 112,
         156, 179, 185, 195
Record of the Day (media monitoring
         service)   95, 172–3
record stores   14, 36, 39, 54, 74, 186–7

recruitment   43–4, 51, 66, 100, 154,
         161, 198
Redhead, Steve   69, 86
redundancy   56, 72, 104, 139, 171,
         197, 202
releases, new music   13, 23, 33–4, 63,
         82, 89, 103–5, 122, 128, 130,
         134, 141–2, 145, 199. *See also*
         marketing campaigns
religious faith. *See* conversion;
         eschatology; millenarianism;
         salvation; spirit; soteriology
research
    academic   14–15, 47, 51, 155–6, 177,
         179 (*see also* research methods)
    market (*see under* market research)
research methods
    in industry   45, 90–1, 169–70, 199–
         200 (*see also* market research)
    limitations of Boltanski and
         Chiapello's   176
    as metaphor for everyday sense-
         making   112–13
    in PMS   159
    in this study   3–4, 44–9, 52, 58, 98,
         106, 191–3
resilience   23, 73, 147
resistance
    to change   26, 91
    to establishment norms   19
    to order and frictionless cultural
         flows   38, 137
    to university marketisation   161
    within the workplace   46, 118, 194
         (*see also* criticism)
Retromania (Simon Reynolds)   28–9
Rough Trade   138–9
royalties   24, 63, 121, 124, 127–8, 131,
         132, 135, 141, 145, 183, 187,
         200, 201, 212–14 n.1

salvation   35, 111. *See also* religious faith
Savage, Mike   35, 65
scene   12, 54–5, 70
scenesters   71, 109
Sedgwick, Eve Kosofsky   46, 49
segmentation, consumer. *See under*
         market research

silences   98, 117–18, 189, 194, 198
single version of truth   128, 135–7, 142
slavery, resonances of in music
        industry   18, 37, 193, 214 n.1
social critique. *See under* critique
social media   56, 79, 83, 90, 100, 104,
        143, 174, 200
Soho (Central London)   70
Sony
    Corporation   56, 58
    DADC distribution   56 (*see also*
        warehousing)
    Music Entertainment   44, 54, 57,
        64–5, 74, 82, 100, 108–9, 127–8,
        160, 212 n.4
    Pictures   45
    Walkman   48, 97
soteriology   111. *See also* religious faith
spirit
    of the age   31–2
    of American capitalism   35
    animal   85
    of capitalism (*see* Weber, Max)
    and capitalist renewal (*see* Boltanski
        and Chiapello)
    entrepreneurial   95
    lost (*see* Stiegler, Bernard)
    of the music industry   66, 129, 161,
        194, 196–8, 202–3
    and personal transformation   35
        (*see also* religious faith)
Spotify   10–11, 29, 36, 39, 73, 107, 110,
        112, 117, 126, 182–3, 193
spreadsheets   3, 128, 137–8, 141–3,
        186, 195
standards
    academic   163, 165, 195
    accounting   25
    cultural   10, 27
    industrial   36–7, 44, 120–3, 134–5,
        137, 141–3, 195
    living   18
statistics   44–5, 66, 91, 110, 169–70, 177,
        180, 181, 194, 197, 200
Sterne, Jonathan   56, 59, 120–1, 135
Stiegler, Bernard (*The Lost Spirit of*
        Capitalism)   28–9
Stiff Records   121

streaming. *See under* platforms
style
    fashion   97–8, 100–1
    life-   32, 70, 108–9, 112, 130
    musical   28, 102
    subcultural 47, 54, 57, 69–71, 74, 87,
        108, 195 (*see also* Hebdige, Dick;
        Thornton, Sarah)
    of thought   35
    visual   168
    of work   42, 141, 168
    writing   58, 79
styling   61–2
Svengalis   11. *See also* record men
Syco   82–4
synchronisation. *See under* licensing

talent pipeline   41, 152, 154. See also
        recruitment
'tech'
    culture   35, 108, 154, 185
    enthusiasm   129
    firms, importance of   13, 29, 58, 75,
        98, 160, 173, 183
    workers   55, 201 (*see also*
        engineering)
theory
    critical (*see* Critical Theory (Frankfurt
        School))
    cultural   158–9, 162, 185
    economic (*see under* economics)
    the end of   49
    game   188
    and practice   158–60, 171, 185–6, 189
Thornton, Sarah   69, 71, 192–3
thought leaders   30, 92, 162, 170, 176,
        182, 185
thresholds
    as analytic object   52
    budgetary   25–6, 124, 128
    organizational   33, 43, 46, 75, 96
    temporal   33
Top 40. *See* charts
trade bodies   2, 4, 14–15, 31, 44, 61,
        107, 193
trade journals
    Billboard   171–2
    Music Ally   63, 175, 182–3

Music Business Worldwide 2, 110, 124, 171, 200, 202
Music Week 66, 100, 161, 171–2, 174
trade unions 17–19, 23, 197, 200–1, 214 n.2
training 23, 27, 32, 41, 51, 72, 99, 152–4, 156–71, 183–90, 192. *See also* graduate training schemes
Travis, Geoff 164. *See also* Rough Trade
Twitter 100, 174, 211 n.1

UK Music 4, 41, 44–5, 60–1, 90, 152–6, 158, 163, 166, 199–200
unbundling 5, 119–22, 140
uncreativity 22, 26, 29, 125
Universal Music Group 9, 33, 44, 57–8, 64, 66, 74, 124, 130, 160, 161

valuation 83, 136, 183
value
    chains 56, 61–2, 122–3, 125, 127–9, 148, 151, 195
    extraction 26, 31, 46, 200–2
    and values 5, 10, 12–16, 21, 38, 42, 57, 65, 67–8, 80–1, 85, 108, 118, 155, 194, 201
vinyl. *See under* formats
Vivendi 58, 74, 90

Walkman. *See under* Sony
Wardle, Ben 150–2, 156
warehousing 20, 24, 26, 27, 54, 56, 127, 142, 144, 148
Warner Music Group 44, 54–5, 57–8, 150, 180
Weber, Max. *See also* bureaucracy
    namedropped by research participants 185

Protestant Ethic and the Spirit of Capitalism, The 12, 16–17
Wikileaks 45–6
Williamson, John 24, 31, 45, 50, 56, 59–60, 124
Wired (magazine) 10, 108. *See also* Anderson, Chris
women. *See also* feminism; gender; misogyny
    in higher education 67–8
    role of in the industry
    in the workforce 20
work
    administrative, office and support 3, 5, 23–4, 26, 95, 97, 120, 123, 131, 138, 148, 152, 212 n.3
    as a conceptual problem 23
    cultural and creative 22, 26–7, 33, 40, 51, 69–71, 97, 112, 198
    ethic (*see under* Weber, Max)
    good 21
    non-creative work 24–5, 133, 200–1 (*see also* uncreativity)
    representations of 18, 23, 81–7, 96, 103
    to rule 118

X Factor. See Idol shows

youthfulness. *See also* age
    in interviewees' justifications, appeals to 72–4, 105–6, 112
    in market imaginaries, perceptions of 22, 92
    in the workforce 41, 43–4, 48, 150, 198–9
YouTube 31, 38, 126